United States
Environmental Protection
Agency

EPA/600/R-11/122
November 2011

I0484118

Plan to Study the Potential Impacts of Hydraulic Fracturing on Drinking Water Resources

Office of Research and Development

US Environmental Protection Agency

Washington, D.C.

November 2011

*Mention of trade names or commercial products does not constitute
endorsement or recommendation for use.*

TABLE OF CONTENTS

LIST OF FIGURES

LIST OF TABLES

LIST OF ACRONYMS AND ABBREVIATIONS

AOE	area of evaluation
API	American Petroleum Institute
ATSDR	Agency for Toxic Substances and Disease Registry
BLM	Bureau of Land Management
CBI	confidential business information
CWT	commercial wastewater treatment facility
DBP	disinfection byproducts
DOE	US Department of Energy
EIA	US Energy Information Administration
EPA	US Environmental Protection Agency
FWS	US Fish and Wildlife Service
GIS	geographic information systems
GWPC	Ground Water Protection Council
mcf/d	thousand cubic feet per day
mg/L	milligram per liter
mmcf/d	million cubic feet per day
NGO	non-governmental organization
NIOSH	National Institute for Occupational Safety and Health
NYS rdSGEIS	New York State Revised Draft Supplemental Generic Environmental Impact Statement
ORD	Office of Research and Development
pCi/L	picocuries per liter
ppmv	parts per million by volume
POTW	publicly owned treatment works
PPRTV	provisional peer-reviewed toxicity value
QA	quality assurance
QAPP	quality assurance project plan
QSAR	quantitative structure-activity relationship
SAB	Science Advisory Board
TDS	total dissolved solids
UIC	underground injection control
USACE	US Army Corps of Engineers
USDW	underground source of drinking water
USGS	US Geological Survey
VOC	volatile organic compound

EXECUTIVE SUMMARY

Natural gas plays a key role in our nation's clean energy future. Recent advances in drilling technologies—including horizontal drilling and hydraulic fracturing—have made vast reserves of natural gas economically recoverable in the US. Responsible development of America's oil and gas resources offers important economic, energy security, and environmental benefits.

Hydraulic fracturing is a well stimulation technique used to maximize production of oil and natural gas in unconventional reservoirs, such as shale, coalbeds, and tight sands. During hydraulic fracturing, specially engineered fluids containing chemical additives and proppant are pumped under high pressure into the well to create and hold open fractures in the formation. These fractures increase the exposed surface area of the rock in the formation and, in turn, stimulate the flow of natural gas or oil to the wellbore. As the use of hydraulic fracturing has increased, so have concerns about its potential environmental and human health impacts. Many concerns about hydraulic fracturing center on potential risks to drinking water resources, although other issues have been raised. In response to public concern, the US Congress directed the US Environmental Protection Agency (EPA) to conduct scientific research to examine the relationship between hydraulic fracturing and drinking water resources.

This study plan represents an important milestone in responding to the direction from Congress. EPA is committed to conducting a study that uses the best available science, independent sources of information, and a transparent, peer-reviewed process that will ensure the validity and accuracy of the results. The Agency will work in consultation with other federal agencies, state and interstate regulatory agencies, industry, non-governmental organizations, and others in the private and public sector in carrying out this study. Stakeholder outreach as the study is being conducted will continue to be a hallmark of our efforts, just as it was during the development of this study plan.

EPA has already conducted extensive stakeholder outreach during the developing of this study plan. The draft version of this study plan was developed in consultation with the stakeholders listed above and underwent a peer review process by EPA's Science Advisory Board (SAB). As part of the review process, the SAB assembled an independent panel of experts to review the draft study plan and to consider comments submitted by stakeholders. The SAB provided EPA with its review of the draft study plan in August 2011. EPA has carefully considered the SAB's recommendations in the development of this final study plan.

The overall purpose of this study is to elucidate the relationship, if any, between hydraulic fracturing and drinking water resources. More specifically, the study has been designed to assess the potential impacts of hydraulic fracturing on drinking water resources and to identify the driving factors that affect the severity and frequency of any impacts. Based on the increasing development of shale gas resources in the US, and the comments EPA received from stakeholders, this study emphasizes hydraulic fracturing in shale formations. Portions of the research, however, are also intended to provide information on hydraulic fracturing in coalbed methane and tight sand reservoirs. The scope of the research includes the hydraulic fracturing water use lifecycle, which is a subset of the greater hydrologic cycle. For the purposes of this study, the hydraulic fracturing water lifecycle begins with water acquisition from surface or ground water and ends with discharge into surface waters or injection into deep wells. Specifically, the water lifecycle for hydraulic fracturing consists of water acquisition, chemical mixing, well injection, flowback and produced water (collectively referred to as "hydraulic fracturing wastewater"), and wastewater treatment and waste disposal.

The EPA study is designed to provide decision-makers and the public with answers to the five fundamental questions associated with the hydraulic fracturing water lifecycle:

- Water Acquisition: What are the potential impacts of large volume water withdrawals from ground and surface waters on drinking water resources?
- Chemical Mixing: What are the possible impacts of surface spills on or near well pads of hydraulic fracturing fluids on drinking water resources?
- Well Injection: What are the possible impacts of the injection and fracturing process on drinking water resources?
- Flowback and Produced Water: What are the possible impacts of surface spills on or near well pads of flowback and produced water on drinking water resources?
- Wastewater Treatment and Waste Disposal: What are the possible impacts of inadequate treatment of hydraulic fracturing wastewaters on drinking water resources?

Answering these questions will involve the efforts of scientists and engineers with a broad range of expertise, including petroleum engineering, fate and transport modeling, ground water hydrology, and toxicology. The study will be conducted by multidisciplinary teams of EPA researchers, in collaboration with outside experts from the public and private sector. The Agency will use existing data from hydraulic fracturing service companies and oil and gas operators, federal and state agencies, and other sources. To supplement this information, EPA will conduct case studies in the field and generalized scenario evaluations using computer modeling. Where applicable, laboratory studies will be conducted to provide a better understanding of hydraulic fracturing fluid and shale rock interactions, the treatability of hydraulic fracturing wastewaters, and the toxicological characteristics of high-priority constituents of concern in hydraulic fracturing fluids and wastewater. EPA has also included a screening analysis of whether hydraulic fracturing activities may be disproportionately occurring in communities with environmental justice concerns.

Existing data will be used answer research questions associated with all stages of the water lifecycle, from water acquisition to wastewater treatment and waste disposal. EPA has requested information from hydraulic fracturing service companies and oil and gas well operators on the sources of water used in hydraulic fracturing fluids, the composition of these fluids, well construction practices, and wastewater treatment practices. EPA will use these data, as well as other publically available data, to help assess the potential impacts of hydraulic fracturing on drinking water resources.

Retrospective case studies will focus on investigating reported instances of drinking water resource contamination in areas where hydraulic fracturing has already occurred. EPA will conduct retrospective case studies at five sites across the US. The sites will be illustrative of the types of problems that have been reported to EPA during stakeholder meetings held in 2010 and 2011. A determination will be made on the presence and extent of drinking water resource contamination as well as whether hydraulic fracturing contributed to the contamination. The retrospective sites will provide EPA with information regarding key factors that may be associated with drinking water contamination.

Prospective case studies will involve sites where hydraulic fracturing will occur after the research is initiated. These case studies allow sampling and characterization of the site before, during, and after water acquisition, drilling, hydraulic fracturing fluid injection, flowback, and gas production. EPA will work with industry and other stakeholders to conduct two prospective case studies in different regions of the US. The data collected during prospective case studies will allow EPA to gain an understanding of hydraulic fracturing practices, evaluate changes in water quality over time, and assess the fate and transport of potential chemical contaminants.

Generalized scenario evaluations will use computer modeling to allow EPA to explore realistic hypothetical scenarios related to hydraulic fracturing activities and to identify scenarios under which hydraulic fracturing activities may adversely impact drinking water resources.

Laboratory studies will be conducted on a limited, opportunistic basis. These studies will often parallel case study investigations. The laboratory work will involve characterization of the chemical and mineralogical properties of shale rock and potentially other media as well as the products that may form after interaction with hydraulic fracturing fluids. Additionally, laboratory studies will be conducted to better understand the treatment of hydraulic fracturing wastewater with respect to fate and transport of flowback or produced water constituents.

Toxicological assessments of chemicals of potential concern will be based primarily on a review of available health effects data. The substances to be investigated include chemicals used in hydraulic fracturing fluids, their degradates and/or reaction products, and naturally occurring substances that may be released or mobilized as a result of hydraulic fracturing. It is not the intent of this study to conduct a complete health assessment of these substances. Where data on chemicals of potential concern are limited, however, quantitative structure-activity relationships—and other approaches—may be used to assess toxicity.

The research projects identified for this study are summarized in Appendix A. EPA is working with other federal agencies to collaborate on some aspects of the research described in this study plan. All research associated with this study will be conducted in accordance with EPA's Quality Assurance Program for environmental data and meet the Office of Research and Development's requirements for the highest level of quality assurance. Quality Assessment Project Plans will be developed, applied, and updated as the research progresses.

A first report of research results will be completed in 2012. This first report will contain a synthesis of EPA's analysis of existing data, available results from retrospective cases studies, and initial results from scenario evaluations, laboratory studies, and toxicological assessments. Certain portions of the work described here, including prospective case studies and laboratory studies, are long-term projects that are not likely to be finished at that time. An additional report in 2014 will synthesize the results of those long-term projects along with the information released in 2012. Figures 10 and 11 summarize the estimated timelines of the research projects outlined in this study plan. EPA is committed to ensuring that the results presented in these reports undergo thorough quality assurance and peer review.

EPA recognizes that the public has raised concerns about hydraulic fracturing that extend beyond the potential impacts on drinking water resources. This includes, for example, air impacts, ecological effects, seismic risks, public safety, and occupational risks. These topics are currently outside the scope of this study plan, but should be examined in the future.

1   INTRODUCTION AND PURPOSE OF STUDY

Hydraulic fracturing is an important means of accessing one of the nation's most vital energy resources, natural gas. Advances in technology, along with economic and energy policy developments, have spurred a dramatic growth in the use of hydraulic fracturing across a wide range of geographic regions and geologic formations in the US for both oil and gas production. As the use of hydraulic fracturing has increased, so have concerns about its potential impact on human health and the environment, especially with regard to possible effects on drinking water resources. These concerns have intensified as hydraulic fracturing has spread from the southern and western regions of the US to other settings, such as the Marcellus Shale, which extends from the southern tier of New York through parts of Pennsylvania, West Virginia, eastern Ohio, and western Maryland. Based on the increasing importance of shale gas as a source of natural gas in the US, and the comments received by EPA from stakeholders, this study plan emphasizes hydraulic fracturing in shale formations containing natural gas. Portions of the research, however, may provide information on hydraulic fracturing in other types of oil and gas reservoirs, such as coalbeds and tight sands.

In response to escalating public concerns and the anticipated growth in oil and natural gas exploration and production, the US Congress directed EPA in fiscal year 2010 to conduct research to examine the relationship between hydraulic fracturing and drinking water resources (US House, 2009):

> *The conferees urge the Agency to carry out a study on the relationship between hydraulic fracturing and drinking water, using a credible approach that relies on the best available science, as well as independent sources of information. The conferees expect the study to be conducted through a transparent, peer-reviewed process that will ensure the validity and accuracy of the data. The Agency shall consult with other federal agencies as well as appropriate state and interstate regulatory agencies in carrying out the study, which should be prepared in accordance with the Agency's quality assurance principles.*

This document presents the final study plan for EPA's research on hydraulic fracturing and drinking water resources, responding to both the direction from Congress and concerns expressed by the public. For this study, EPA defines "drinking water resources" to be any body of water, ground or surface, that could currently, or in the future, serve as a source of drinking water for public or private water supplies.

The overarching goal of this research is to answer the following questions:

- Can hydraulic fracturing impact drinking water resources?
- If so, what conditions are associated with these potential impacts?

To answer these questions, EPA has identified a set of research activities associated with each stage of the hydraulic fracturing water lifecycle (Figure 1), from water acquisition through the mixing of chemicals and actual fracturing to post-fracturing production, including the management of hydraulic fracturing wastewaters (commonly referred to as "flowback" and "produced water") and ultimate

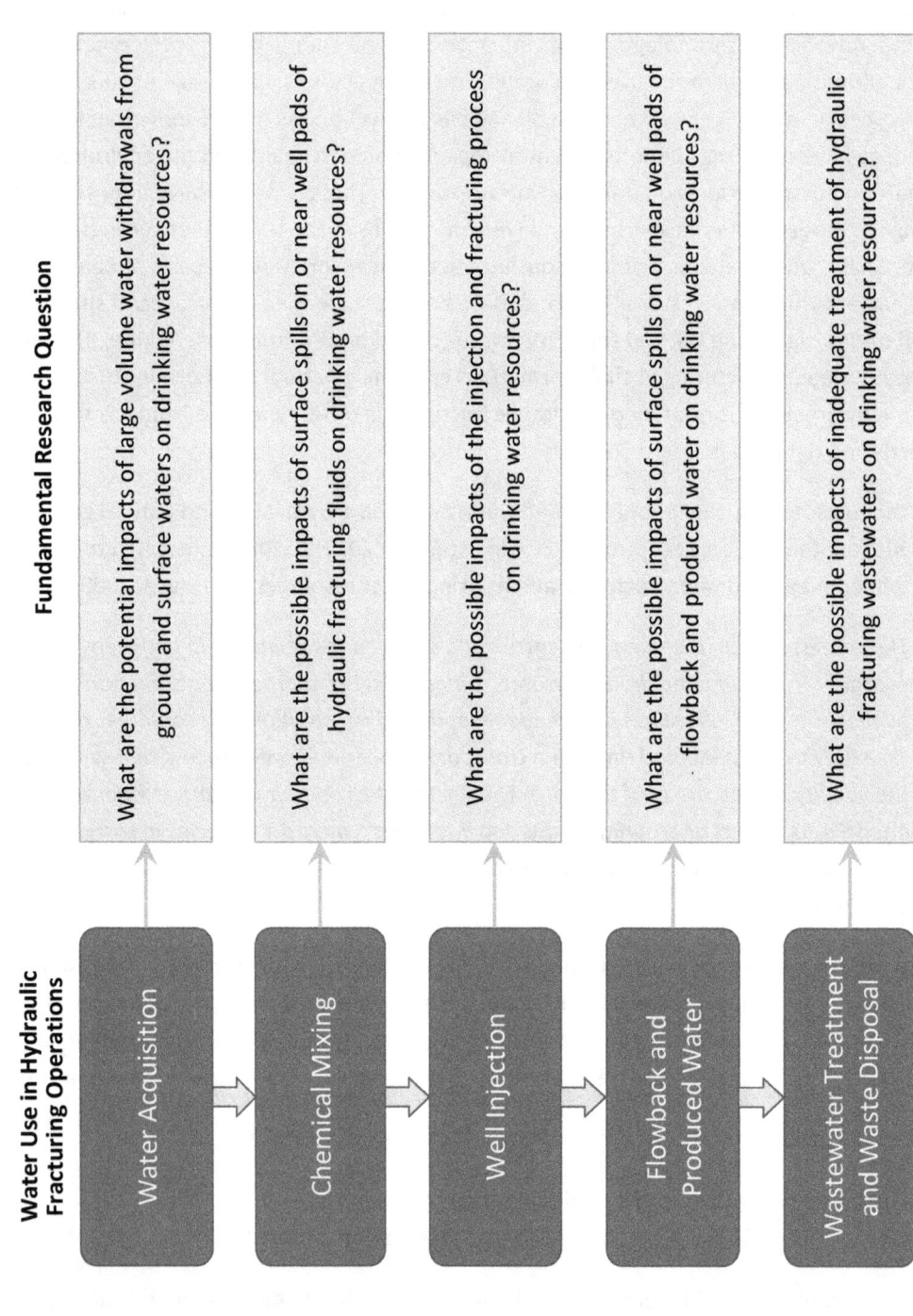

FIGURE 1. FUNDAMENTAL RESEARCH QUESTIONS POSED FOR EACH IDENTIFIED STAGE

treatment and disposal. These research activities will identify potential impacts to drinking water resources of water withdrawals as well as fate and transport of chemicals associated with hydraulic fracturing. Information about the toxicity of contaminants of concern will also be gathered. This information can then be used to assess the potential risks to drinking water resources from hydraulic fracturing activities. Ultimately, the results of this study will inform the public and provide policymakers at all levels with sound scientific knowledge that can be used in decision-making processes.

The study plan is organized as follows:

- Chapter 2 details the process for developing the study plan and the criteria for prioritizing the research.
- Chapter 3 provides a brief overview of unconventional oil and natural gas resources and production.
- Chapter 4 outlines the hydraulic fracturing water lifecycle and the research questions associated with each stage of the lifecycle.
- Chapter 5 briefly describes the research approach.
- Chapter 6 provides background information on each stage of the hydraulic fracturing water lifecycle and describes research specific to each stage.
- Chapter 7 provides background information and describes research to assess concerns pertaining to environmental justice.
- Chapter 8 describes how EPA is collecting, evaluating, and analyzing existing data.
- Chapter 9 presents the retrospective and prospective case studies.
- Chapter 10 discusses scenario evaluations and modeling using existing data and new data collected from case studies.
- Chapter 11 explains how EPA will characterize toxicity of constituents associated with hydraulic fracturing fluids to human health.
- Chapter 12 summarizes how the studies will address the research questions posed for each stage of the water lifecycle.
- Chapter 13 notes additional areas of concern relating to hydraulic fracturing that are currently outside the scope of this study plan.

Also included at the end of this document are eight appendices and a glossary.

2 PROCESS FOR STUDY PLAN DEVELOPMENT

2.1 STAKEHOLDER INPUT

Stakeholder input played an important role in the development of the hydraulic fracturing study plan. Many opportunities were provided for the public to comment on the study scope and case study locations. The study plan was informed by information exchanges involving experts from the public and private sectors on a wide range of technical issues. EPA will continue to engage stakeholders throughout the course of the study and as results become available.

EPA has engaged stakeholders in the following ways:

Federal, state, and tribal partner consultations. Webinars were held with state partners in May 2010, with federal partners in June 2010, and with Indian tribes in August 2010. The state webinar included representatives from 21 states as well as representatives from the Association of State Drinking Water Administrators, the Association of State and Interstate Water Pollution Control Administrators, the Ground Water Protection Council (GWPC), and the Interstate Oil and Gas Compact Commission. Federal partners included the Bureau of Land Management (BLM), the US Geological Survey (USGS), the US Fish and Wildlife Service (FWS), the US Forest Service, the US Department of Energy (DOE), the US Army Corps of Engineers (USACE), the National Park Service, and the Agency for Toxic Substances and Disease Registry (ATSDR). There were 36 registered participants for the tribal webinar, representing 25 tribal governments. In addition, a meeting with the Haudenosaunee Environmental Task Force in August 2010 included 20 representatives from the Onondaga, Mohawk, Tuscarora, Cayuga, and Tonawanda Seneca Nations. The purpose of these consultations was to discuss the study scope, data gaps, opportunities for sharing data and conducting joint studies, and current policies and practices for protecting drinking water resources.

Sector-specific meetings. Separate webinars were held in June 2010 with representatives from industry and non-governmental organizations (NGOs) to discuss the public engagement process, the scope of the study, coordination of data sharing, and other key issues. Overall, 176 people representing various natural gas production and service companies and industry associations participated in the webinars, as well as 64 people representing NGOs.

Informational public meetings. Public information meetings were held between July and September 2010 in Fort Worth, Texas; Denver, Colorado; Canonsburg, Pennsylvania; and Binghamton, New York. At these meetings, EPA presented information on its reasons for studying hydraulic fracturing, an overview of what the study might include, and how stakeholders can be involved. Opportunities to present oral and written comments were provided, and EPA specifically asked for input on the following questions:

- What should be EPA's highest priorities?
- Where are the gaps in current knowledge?
- Are there data and information EPA should know about?
- Where do you recommend EPA conduct case studies?

Total attendance for all of the informational public meetings exceeded 3,500 and more than 700 verbal comments were heard.

Summaries of the stakeholder meetings can be found at http://www.epa.gov/hydraulicfracturing.

Technical Workshops. Technical workshops organized by EPA were in February and March 2011 to explore the following focus areas: Chemical and Analytical Methods (February 24-25), Well Construction and Operations (March 10-11), Fate and Transport (March 28-29), and Water Resource Management (March 29-30). The technical workshops centered around three goals: (1) inform EPA of the current technology and practices being used in hydraulic fracturing; (2) identify existing/current research related to the potential impacts of hydraulic fracturing on drinking water resources; and (3) provide an opportunity for EPA scientists to interact with technical experts. EPA invited technical experts from the oil and natural gas industry, consulting firms, laboratories, state and federal agencies, and environmental organizations to participate in the workshops. The information presented at the workshops will inform the research outlined in this study plan.

Other opportunities to comment. In addition to conducting the meetings listed above, EPA provided stakeholders with opportunities to submit electronic or written comments on the hydraulic fracturing study. EPA received over 5,000 comments, which are summarized in Appendix B.

2.2   SCIENCE ADVISORY BOARD INVOLVEMENT

The EPA Science Advisory Board (SAB) is a federal advisory committee that provides a balanced, expert assessment of scientific matters relevant to EPA. An important function of the SAB is to review EPA's technical programs and research plans. Members of the advisory board and *ad hoc* panels are nominated by the public and are selected based on factors such as technical expertise, knowledge, and experience. The panel formation process, which is designed to ensure public transparency, also includes an assessment of potential conflicts of interest or lack of impartiality. SAB panels are composed of individuals with a wide range of expertise to ensure that the technical advice is comprehensive and balanced.

EPA's Office of Research and Development (ORD) has engaged the SAB through the development of this study plan. This process is described below.

Initial SAB review of the study plan scope. During fiscal year 2010, ORD developed a document that presented the scope and initial design of the study (USEPA, 2010a). The document was submitted to the SAB's Environmental Engineering Committee for review in March 2010. In its response to EPA in June 2010 (USEPA, 2010c), the SAB recommended that:

- Initial research should be focused on potential impacts to drinking water resources, with later research investigating more general impacts on water resources.
- Engagement with stakeholders should occur throughout the research process.
- Five to ten in-depth case studies at "locations selected to represent the full range of regional variability of hydraulic fracturing across the nation" should be part of the research plan.

EPA concurred with these recommendations and developed the draft study plan accordingly.

The SAB also cautioned EPA against studying all aspects of oil and gas production, stating that the study should "emphasize human health and environmental concerns specific to, or significantly influenced by, hydraulic fracturing rather than on concerns common to all oil and gas production activities." Following this advice, EPA focused the draft study plan on features of oil and gas production that are particular to—or closely associated with—hydraulic fracturing, and their impacts on drinking water resources.

SAB review of the draft study plan. EPA developed a *Draft Plan to Study the Potential Impacts of Hydraulic Fracturing on Drinking Water Resources* (USEPA, 2011a) after receiving the SAB's review of the scoping document in June 2010 and presented the draft plan to the SAB for review in February 2011. The SAB formed a panel to review the plan,[1] which met in March 2011. The panel developed an initial review of the draft study plan and subsequently held two public teleconference calls in May 2011 to discuss this review. The review panel's report was discussed by the full SAB during a public teleconference in July 2011. The public had the opportunity to submit oral and written comments at each meeting and teleconference of the SAB. As part of the review process, the public submitted over 300 comments for consideration.[2] The SAB considered the comments submitted by the public as they formulated their review of the draft study plan. In their final report to the Agency, the SAB generally supported the research approach outlined in the draft study plan and agreed with EPA's use of the water lifecycle as a framework for the study (EPA, 2011b). EPA carefully considered and responded to the SAB's recommendations on September 27, 2011.[3]

2.3 RESEARCH PRIORITIZATION

In developing this study plan, EPA considered the results of a review of the literature,[4] technical workshops, comments received from stakeholders, and input from meetings with interested parties, including other federal agencies, Indian tribes, state agencies, industry, and NGOs. EPA also considered recommendations from the SAB reviews of the study plan scope (USEPA, 2010c) and the draft study plan (USEPA, 2011b).

In response to the request from Congress, EPA identified fundamental questions (see Figure 1) that frame the scientific research to evaluate the potential for hydraulic fracturing to impact drinking water resources. Following guidance from the SAB, EPA used a risk-based prioritization approach to identify research that addresses the most significant potential risks at each stage of the hydraulic fracturing water lifecycle. The risk assessment paradigm (i.e., exposure assessment, hazard identification, dose-response relationship assessment, and risk characterization) provides a useful framework for asking scientific questions and focusing research to accomplish the stated goals of this study, as well as to inform full risk assessments in the future. For the current study, emphasis is placed on exposure assessment and hazard identification. Exposure assessment will be informed by work on several tasks including, but not limited to, modeling (i.e., water acquisition, injection/flowback/production, wastewater management), case studies, and evaluation of existing data. Analysis of the chemicals used in hydraulic fracturing, how they are used, and their fate will provide useful data for hazard identification. A definitive evaluation of dose-response relationships and a comprehensive risk characterization are beyond the scope of this study.

---

[1] Biographies on the members of the SAB panel can be found at http://yosemite.epa.gov/sab/sabproduct.nsf/ fedrgstr_activites/HFSP!OpenDocument&TableRow=2.1#2.

[2] These comments are available as part of the material from the SAB public meetings, and can be found at http://yosemite.epa.gov/sab/SABPRODUCT.NSF/81e39f4c09954fcb85256ead006be86e/ d3483ab445ae61418525775900603e79!OpenDocument&TableRow=2.2#2.

[3] See http://yosemite.epa.gov/sab/sabproduct.nsf/2BC3CD632FCC0E99852578E2006DF890/$File/EPA-SAB-11-012_Response_09-27-2011.pdf and http://water.epa.gov/type/groundwater/uic/class2/hydraulicfracturing/ upload/final_epa_response_to_sab_review_table_091511.pdf.

[4] The literature review includes information from more than 120 articles, reports, presentations and other materials. Information resulting from this literature review is incorporated throughout this study plan.

Other criteria considered in prioritizing research activities included:

- *Relevance:* Only work that may directly inform an assessment of the potential impacts of hydraulic fracturing on drinking water resources was considered.
- *Precedence:* Work that needs to be completed before other work can be initiated received a higher priority.
- *Uniqueness of the contribution:* Relevant work already underway by others received a lower priority for investment by EPA.
- *Funding:* Work that could provide EPA with relevant results given a reasonable amount of funding received a higher priority.
- *Leverage:* Relevant work that EPA could leverage with outside investigators received a higher priority.

As the research progresses, EPA may determine that modifying the research approach outlined in this study plan or conducting additional research within the overall scope of the plan is prudent in order to better answer the research questions. In that case, modifications to the activities that are currently planned may be necessary.

2.4 NEXT STEPS

EPA is committed to continuing our extensive outreach efforts to stakeholder as the study progresses. This will include:

- Periodic updates will be provided to the public on the progress of the research.
- A peer-reviewed study report providing up-to-date research results will be released to the public in 2012.
- A second, peer-reviewed study report will be released to the public in 2014. This report will include information from the entire body of research described in this study plan.

2.5 INTERAGENCY COOPERATION

In a series of meetings, EPA consulted with several federal agencies regarding research related to hydraulic fracturing. EPA met with representatives from DOE[5] and DOE's National Energy Technology Laboratory, USGS, and USACE to learn about research that those agencies are involved in and to identify opportunities for collaboration and leverage. As a result of those meetings, EPA has identified work being done by others that can inform its own study on hydraulic fracturing. EPA and other agencies are collaborating on information gathering and research efforts. In particular, the Agency is coordinating with DOE and USGS on existing and future research projects relating to hydraulic fracturing. Meetings between EPA and DOE have enabled the sharing of each agency's research on hydraulic fracturing and the exchange of information among experts.

---

[5] DOE's efforts are briefly summarized in Appendix C.

Specifically, DOE, USGS, USACE, and the Pennsylvania Geological Survey have committed to collaborate with EPA on this study. All four are working with EPA on one of the prospective case studies (Washington County, Pennsylvania). USGS is performing stable isotope analysis of strontium for all retrospective and prospective case studies. USGS is also sharing data on their studies in Colorado and New Mexico.

Federal agencies also had an opportunity to provide comments on EPA's *Draft Plan to Study the Potential Impacts of Hydraulic Fracturing on Drinking Water Resources* through an interagency review. EPA received comments from the ATSDR, DOE, BLM, USGS, FWS, the Office of Management and Budget, the US Energy Information Administration (EIA), the Occupational Safety and Health Administration, and the National Institute of Occupational Safety and Health (NIOSH). These comments were reviewed and the study plan was appropriately modified.

2.6 QUALITY ASSURANCE

All EPA-funded intramural and extramural research projects that generate or use environmental data to make conclusions or recommendations must comply with Agency Quality Assurance (QA) Program requirements (USEPA, 2002). EPA recognizes the value of using a graded approach such that QA requirements are based on the importance of the work to which the program applies. Given the significant national interest in the results of this study, the following rigorous QA approach will be used:

- Research projects will comply with Agency requirements and guidance for quality assurance project plans (QAPPs), including the use of systematic planning.
- Technical systems audits, audits of data quality, and data usability (quality) assessments will be conducted as described in QAPPs.
- Performance evaluations of analytical systems will be conducted.
- Products[6] will undergo QA review.
- Reports will have readily identifiable QA sections.
- Research records will be managed according to EPA's record schedule 501 for *Applied and Directed Scientific Research* (USEPA, 2009).

All EPA organizations involved with the generation or use of environmental data are supported by QA professionals who oversee the implementation of the QA program for their organization. Given the cross-organizational nature of the research, EPA has identified a Program QA Manager who will coordinate the rigorous QA approach described above and oversee its implementation across all participating organizations. The organizational complexity of the hydraulic fracturing research effort also demands that a quality management plan be written to define the QA-related policies, procedures, roles, responsibilities, and authorities for this research. The plan will document consistent QA procedures and practices that may otherwise vary between organizations.

---

[6] Applicable products may include reports, journal articles, symposium/conference papers, extended abstracts, computer products/software/models/databases and scientific data.

3   OVERVIEW OF UNCONVENTIONAL OIL AND NATURAL GAS PRODUCTION

Hydraulic fracturing is often used to stimulate the production of hydrocarbons from unconventional oil and gas reservoirs, which include shales, coalbeds, and tight sands.[7] "Unconventional reservoirs" refers to oil and gas reservoirs whose porosity, permeability, or other characteristics differ from those of conventional sandstone and carbonate reservoirs (USEIA, 2011a). Many of these formations have poor permeability, so reservoir stimulation techniques such as hydraulic fracturing are needed to make oil and gas production cost-effective. In contrast, conventional oil and gas reservoirs have a higher permeability and operators generally have not used hydraulic fracturing. However, hydraulic fracturing has become increasingly used to increase the gas flow in wells that are considered conventional reservoirs and make them even more economically viable (Martin and Valkó, 2007).

Unconventional natural gas development has become an increasingly important source of natural gas in the US in recent years. It accounted for 28 percent of total natural gas production in 1998 (Arthur et al., 2008). Figure 2 illustrates that this percentage rose to 50 percent in 2009, and is projected to increase to 60 percent in 2035 (USEIA, 2010).

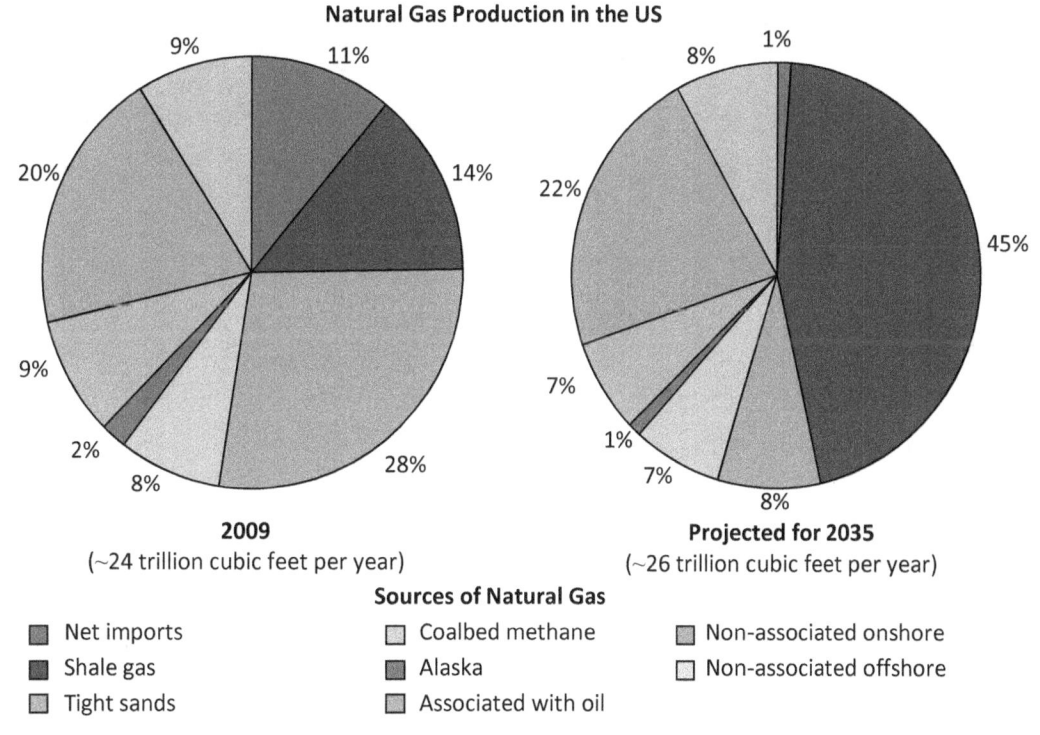

FIGURE 2. NATURAL GAS PRODUCTION IN THE US (DATA FROM USEIA, 2010)

---

[7] Hydraulic fracturing has also been used for other purposes, such as removing contaminants from soil and ground water at waste disposal sites, making geothermal wells more productive, and completing water wells (Nemat-Nassar et al., 1983; New Hampshire Department of Environmental Services, 2010).

This rise in hydraulic fracturing activities to produce gas from unconventional reservoirs is also reflected in the number of drilling rigs operating in the US. There were 603 horizontal gas rigs in June 2010, an increase of 277 from the previous year (Baker Hughes, 2010). Horizontal rigs are commonly used when hydraulic fracturing is used to stimulate gas production from shale formations.

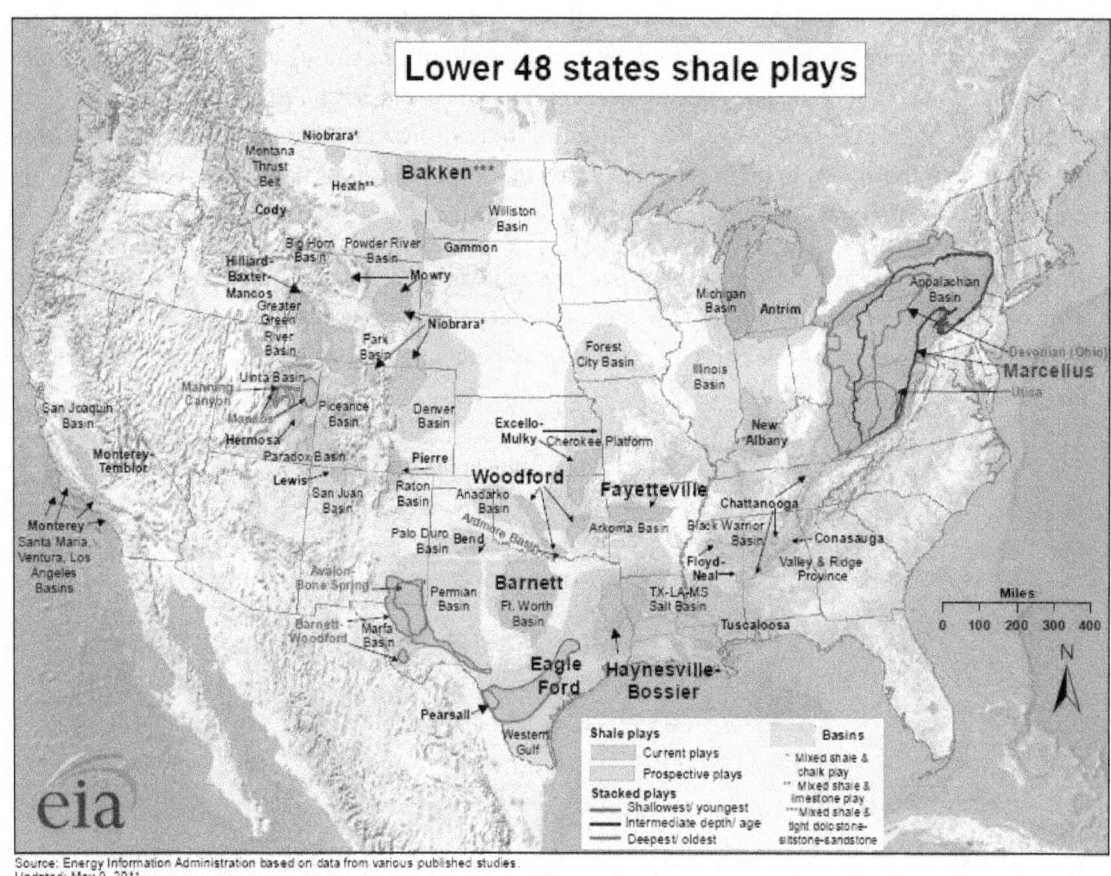

FIGURE 3. SHALE GAS PLAYS IN THE CONTIGUOUS US

Shale gas extraction. Shale rock formations have become an important source of natural gas in the US and can be found in many locations across the country, as shown in Figure 3. Depths for shale gas formations can range from 500 to 13,500 feet below the earth's surface (GWPC and ALL Consulting, 2009). At the end of 2009, the five most productive shale gas fields in the country—the Barnett, Haynesville, Fayetteville, Woodford, and Marcellus Shales—were producing 8.3 billion cubic feet of natural gas per day (Zoback et al., 2010). According to recent figures from EIA, shale gas constituted 14 percent of the total US natural gas supply in 2009, and will make up 45 percent of the US gas supply in 2035 if current trends and policies persist (USEIA, 2010).

Oil production has similarly increased in oil-bearing shales following the increased use of hydraulic fracturing. Proven oil production from shales has been concentrated primarily in the Williston Basin in North Dakota, although oil production is increasing in the Eagle Ford Shale in Texas, the Niobrara Shale in Colorado, Nebraska, and Wyoming, and the Utica Shale in Ohio (USEIA, 2010, 2011b; OilShaleGas.com, 2010).

Production of coalbed methane. Coalbed methane is formed as part of the geological process of coal generation and is contained in varying quantities within all coal. Depths of coalbed methane formations range from 450 feet to greater than 10,000 feet (Rogers et al., 2007; National Research Council, 2010). At greater depths, however, the permeability decreases and production is lower. Below 7,000 feet, efficient production of coalbed methane can be challenging from a cost-effectiveness perspective (Rogers et al., 2007). Figure 4 displays coalbed methane reservoirs in the contiguous US. In 1984, there were very few coalbed methane wells in the US; by 1990, there were almost 8,000, and in 2000, there were almost 14,000 (USEPA, 2004). In 2009, natural gas production from coalbed methane reservoirs made up 8 percent of the total US natural gas production; this percentage is expected to remain relatively constant over the next 20 years if current trends and policies persist (USEIA, 2010). Production of gas from coalbeds almost always requires hydraulic fracturing (USEPA, 2004), and many existing coalbed methane wells that have not been fractured are now being considered for hydraulic fracturing.

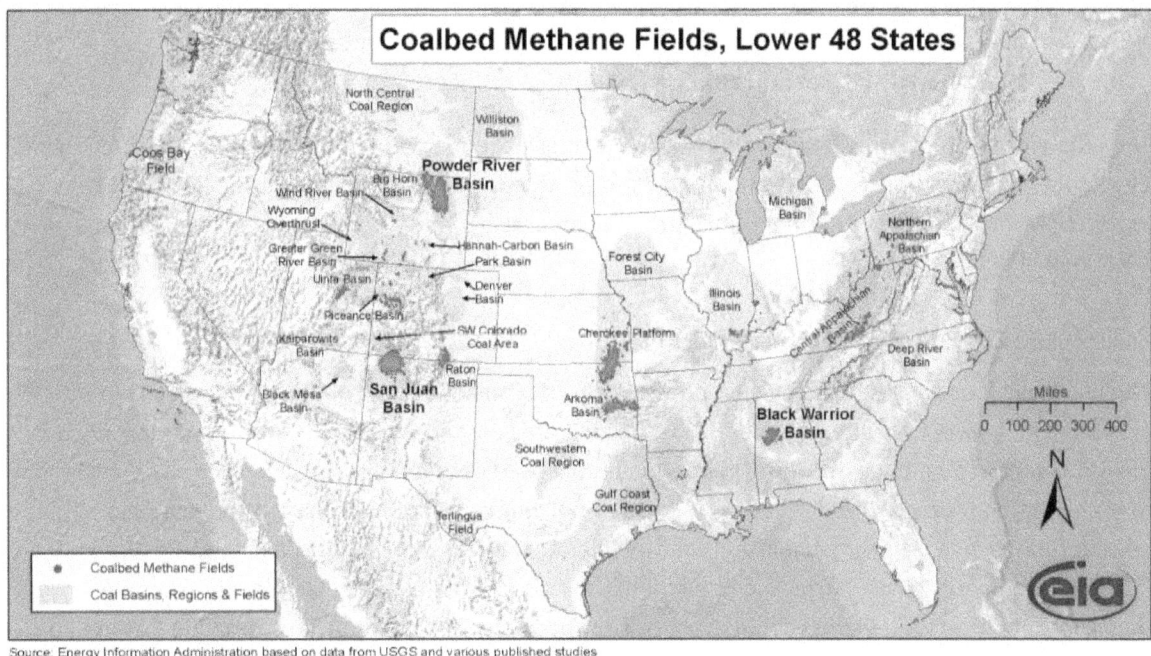

Source: Energy Information Administration based on data from USGS and various published studies
Updated: April 8, 2009

FIGURE 4. COALBED METHANE DEPOSITS IN THE CONTIGUOUS US

Tight sands. Tight sands (gas-bearing, fine-grained sandstones or carbonates with a low permeability) accounted for 28 percent of total gas production in the US in 2009 (USEIA, 2010), but may account for as much as 35 percent of the nation's recoverable gas reserves (Oil and Gas Investor, 2005). Figure 5 shows the locations of tight gas plays in the US. Typical depths of tight sand formations range from 1,200 to 20,000 feet across the US (Prouty, 2001). Almost all tight sand reservoirs require hydraulic fracturing to release gas unless natural fractures are present.

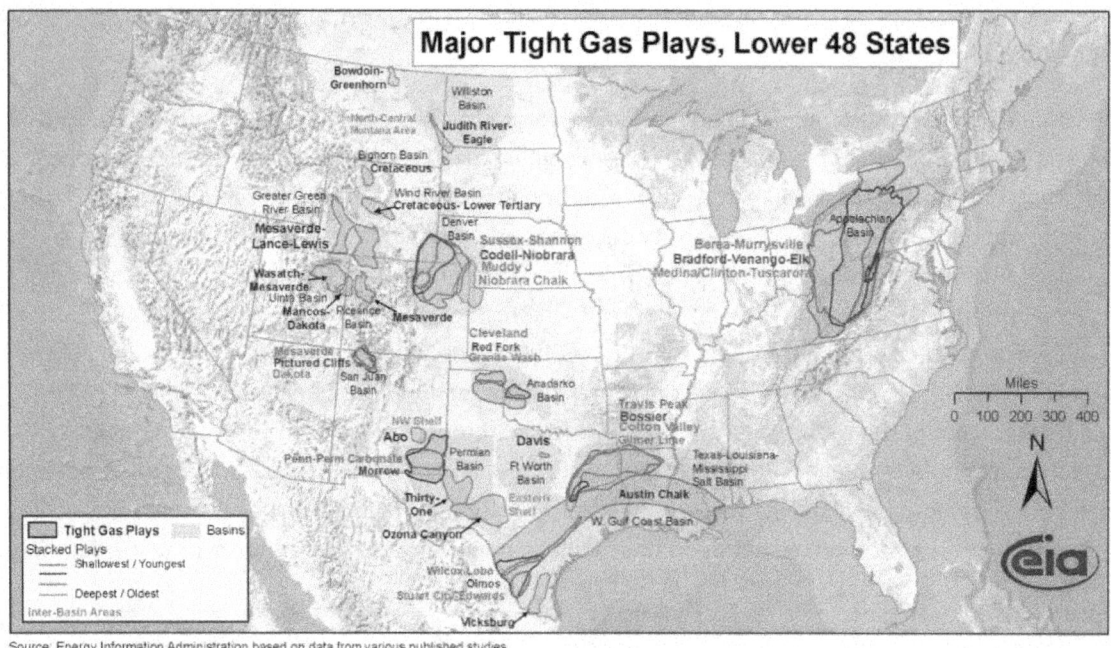

Source: Energy Information Administration based on data from various published studies
Updated: June 6, 2010

FIGURE 5. MAJOR TIGHT GAS PLAYS IN THE CONTIGUOUS US

The following sections provide an overview of how site selection and preparation, well construction and development, hydraulic fracturing, and natural gas production apply to unconventional natural gas production. The current regulatory framework that governs hydraulic fracturing activities is briefly described in Section 3.5.

3.1 SITE SELECTION AND PREPARATION

The hydraulic fracturing process begins with exploring possible well sites, followed by selecting and preparing an appropriate site. In general, appropriate sites are those that are considered most likely to yield substantial quantities of natural gas at minimum cost. Other factors, however, may be considered in the selection process. These include proximity to buildings and other infrastructure, geologic considerations, and proximity to natural gas pipelines or the feasibility of installing new pipelines (Chesapeake Energy, 2009). Laws and regulations may also influence site selection. For example, applicants applying for a Marcellus Shale natural gas permit in Pennsylvania must provide information about proximity to coal seams and distances from surface waters and water supplies (PADEP, 2010a).

During site preparation, an area is cleared to provide space to accommodate one or more wellheads; tanks and/or pits for holding water, used drilling fluids, and other materials; and space for trucks and other equipment. At a typical shale gas production site, a 3- to 5-acre space is needed in addition to access roads for transporting materials to and from the well site. If not already present, both the site and access roads need to be built or improved to support heavy equipment.

3.2 WELL CONSTRUCTION AND DEVELOPMENT

3.2.1 TYPES OF WELLS

Current practices in drilling for natural gas include drilling vertical, horizontal, and directional (S-shaped) wells. On the following pages, two different well completions are depicted with one in a typical deep shale gas-bearing formation like the Marcellus Shale (Figure 6) and one in a shallower environment (Figure 7), which is often encountered where coalbed methane or tight sand gas production takes place.

The figures demonstrate a significant difference in the challenges posed for protecting underground drinking water resources. The deep shale gas environment typically has several thousand feet of rock formation separating underground drinking water resources, while the other shows that gas production can take place at shallow depths that also contain underground sources of drinking water (USDWs). The water well in Figure 7 illustrates an example of the relative depths of a gas well and a water well.

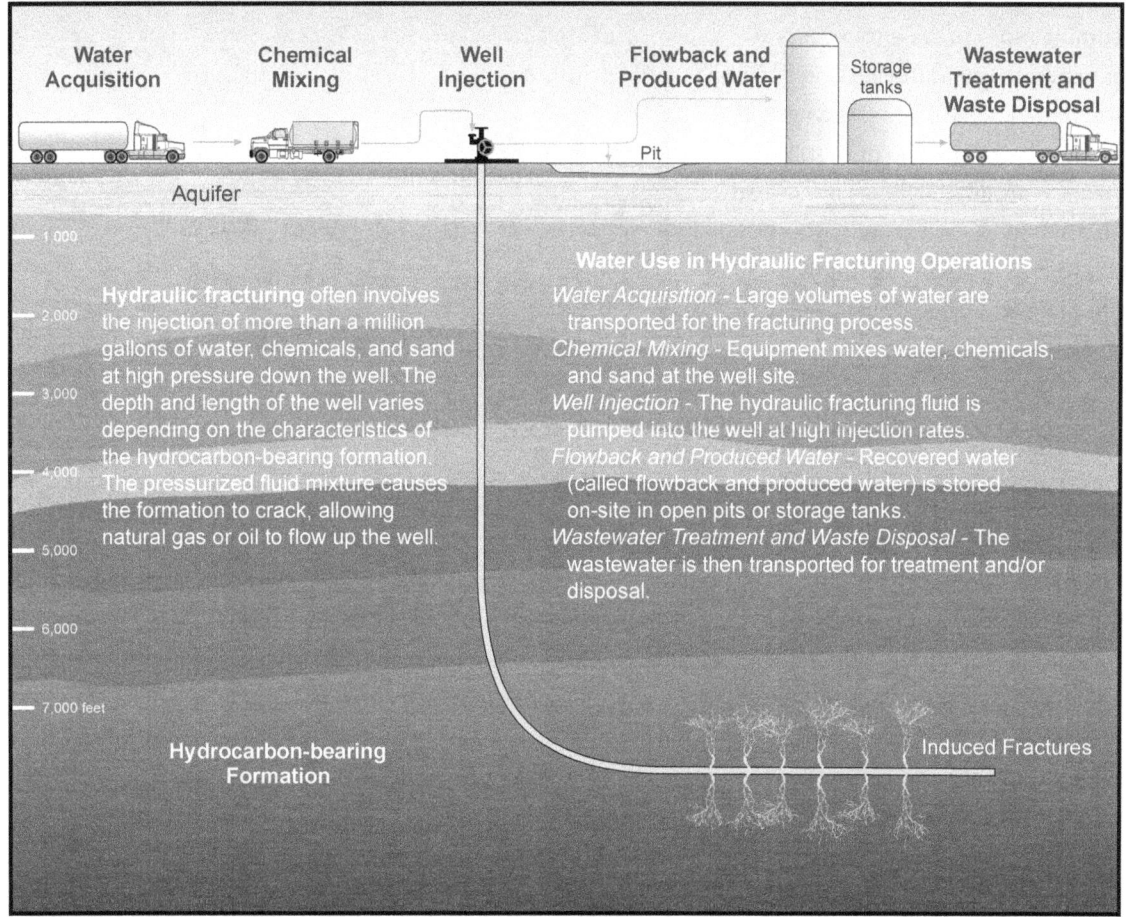

FIGURE 6. ILLUSTRATION OF A HORIZONTAL WELL SHOWING THE WATER LIFECYCLE IN HYDRAULIC FRACTURING

Figure 6 depicts a horizontal well, which is composed of both vertical and horizontal legs. The depth and length of the well varies with the location and properties of the gas-containing formation. In unconventional cases, the well can extend more than a mile below the ground surface (Chesapeake

Energy, 2010) while the "toe" of the horizontal leg can be almost two miles from the vertical leg (Zoback et al., 2010). Horizontal drilling provides more exposure to a formation than a vertical well does, making gas production more economical. It may also have the advantage of limiting environmental disturbances on the surface because fewer wells are needed to access the natural gas resources in a particular area (GWPC and ALL Consulting, 2009).

The technique of multilateral drilling is becoming more prevalent in gas production in the Marcellus Shale region (Kargbo et al., 2010) and elsewhere. In multilateral drilling, two or more horizontal production holes are drilled from a single surface location (Ruszka, 2007) to create an arrangement resembling an upside-down tree, with the vertical portion of the well as the "trunk," and multiple "branches" extending out from it in different directions and at different depths.

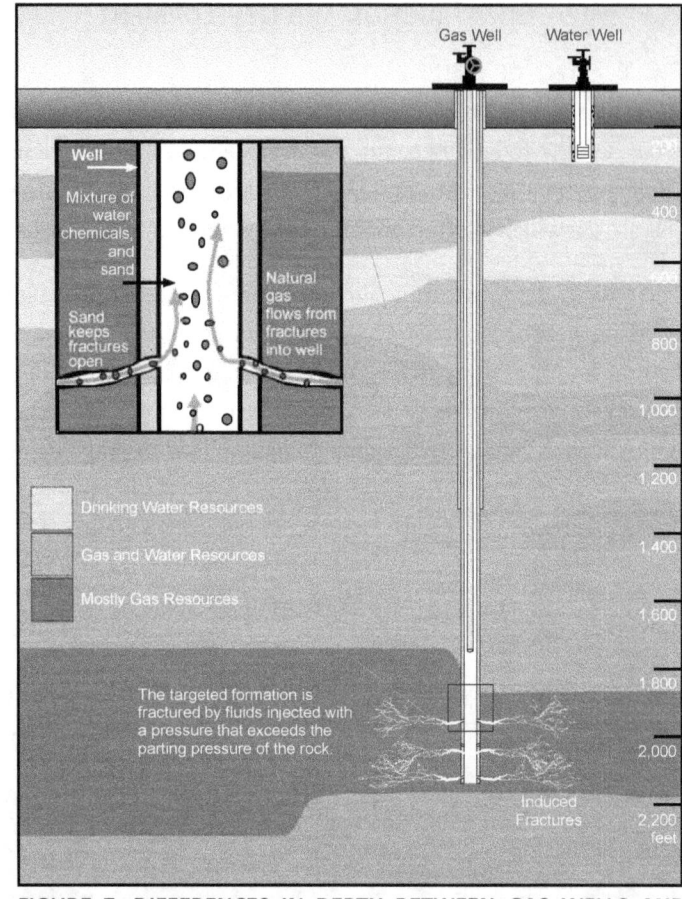

FIGURE 7. DIFFERENCES IN DEPTH BETWEEN GAS WELLS AND DRINKING WATER WELLS

3.2.2 WELL DESIGN AND CONSTRUCTION

According to American Petroleum Institute (API, 2009a), the goal of well design is to "ensure the environmentally sound, safe production of hydrocarbons by containing them inside the well, protecting ground water resources, isolating the production formations from other formations, and by proper execution of hydraulic fractures and other stimulation operations." Proper well construction is essential for isolating the production zone from drinking water resources, and includes drilling a hole, installing steel pipe (casing), and cementing the pipe in place. These activities are repeated multiple times throughout the drilling event until the well is completed.

Drilling. A drilling string—composed of a drill bit, drill collars, and a drill pipe—is used to drill the well. During the drilling process, a drilling fluid such as compressed air or a water- or oil-based liquid ("mud") is circulated down the drilling string. Water-based liquids typically contain a mixture of water, barite, clay, and chemical additives (OilGasGlossary.com, 2010). Drilling fluid serves multiple purposes, including cooling the drill bit, lubricating the drilling assembly, removing the formation cuttings,

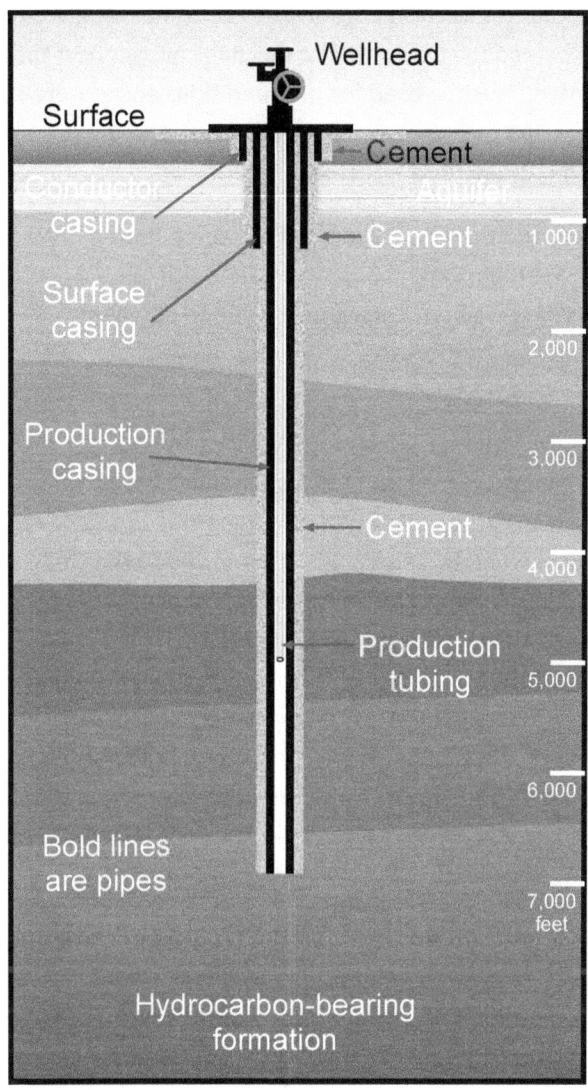

FIGURE 8. WELL CONSTRUCTION

maintaining the pressure control of the well, and stabilizing the hole being drilled. Once removed from the wellbore, both drilling liquids and drill cuttings must be treated, recycled, and/or disposed.

Casing. Casings are steel pipes that line the borehole and serve to isolate the geologic formation from the materials and equipment in the well. The casing also prevents the borehole from caving in, confines the injected/produced fluid to the wellbore and the intended production zone, and provides a method of pressure control. Thus, the casing must be capable of withstanding the external and internal pressures encountered during the installation, cementing, fracturing, and operation of the well. When fluid is confined within the casing, the possibility of contamination of zones adjacent to the well is greatly diminished. In situations where the geologic formation is considered competent and will not collapse upon itself, an operator may choose to forego casing in what is called an open hole completion.

Figure 8 illustrates the different types of casings that may be used in well construction: conductor, surface, intermediate (not shown), and production. Each casing serves a unique purpose. Ideally, the surface casing should extend below the base of the deepest USDW and be cemented to the surface. This casing isolates the USDW and provides protection from contamination during drilling, completion, and operation of the well. Note that the shallow portions of the well may have multiple layers of casing and cement, isolating the production area from the surrounding formation. For each casing, a hole is drilled and the casing is installed and cemented into place.

Casings should be positioned in the center of the borehole using casing centralizers, which attach to the outside of the casing. A centralized casing improves the likelihood that it will be completely surrounded by cement during the cementing process, leading to the effective isolation of the well from USDWs. The number, depth, and cementing of the casings required varies and is set by the states.

Cementing. Once the casing is inserted in the borehole, it is cemented into place by pumping cement slurry down the casing and up the annular space between the formation and the outside of the casing.

15

The principal functions of the cement (for vertical wells or the vertical portion of a horizontal well) are to act as a barrier to migration of fluids up the wellbore behind the casing and to mechanically support the casing. To accomplish these functions, the proper cement must be used for the conditions encountered in the borehole. Additionally, placement of the cement and the type of cement used in the well must be carefully planned and executed to ensure that the cement functions effectively.

The presence of the cement sheath around each casing and the effectiveness of the cement in preventing fluid movement are the major factors in establishing and maintaining the mechanical integrity of the well, although even a correctly constructed well can fail over time due to downhole stresses and corrosion (Bellabarba et al., 2008).

3.3   HYDRAULIC FRACTURING

After the well is constructed, the targeted formation (shale, coalbed, or tight sands) is hydraulically fractured to stimulate natural gas production. As noted in Figure 6, the hydraulic fracturing process requires large volumes of water that must be withdrawn from the source and transported to the well site. Once on site, the water is mixed with chemicals and a propping agent (called a proppant). Proppants are solid materials that are used to keep the fractures open after pressure is reduced in the well. The most common proppant is sand (Carter et al., 1996), although resin-coated sand, bauxite, and ceramics have also been used (Arthur et al., 2008; Palisch et al., 2008). Most, if not all, water-based fracturing techniques use proppants. There are, however, some fracturing techniques that do not use proppants. For example, nitrogen gas is commonly used to fracture coalbeds and does not require the use of proppants (Rowan, 2009).

After the production casing has been perforated by explosive charges introduced into the well, the rock formation is fractured when hydraulic fracturing fluid is pumped down the well under high pressure. The fluid is also used to carry proppant into the targeted formation and enhance the fractures. As the injection pressure is reduced, recoverable fluid is returned to the surface, leaving the proppant behind to keep the fractures open. The inset in Figure 7 illustrates how the resulting fractures create pathways in otherwise impermeable gas-containing formations, resulting in gas flow to the well for production.

The fluid that returns to the surface can be referred to as either "flowback" or "produced water," and may contain both hydraulic fracturing fluid and natural formation water. "Flowback" can be considered a subset of "produced water." However, for this study, EPA considers "flowback" to be the fluid returned to the surface after hydraulic fracturing has occurred, but before the well is placed into production, while "produced water" is the fluid returned to the surface after the well has been placed into production. In this study plan, flowback and produced water are collectively referred to as "hydraulic fracturing wastewaters." These wastewaters are typically stored on-site in tanks or pits before being transported for treatment, disposal, land application, and/or discharge. In some cases, flowback and produced waters are treated to enable the recycling of these fluids for use in hydraulic fracturing.

3.4 WELL PRODUCTION AND CLOSURE

Natural gas production rates can vary between basins as well as within a basin, depending on geologic factors and completion techniques. For example, the average well production rates for coalbed methane formations range from 50 to 500 thousand cubic feet per day (mcf/d) across the US, with maximum production rates reaching 20 million cubic feet per day (mmcf/d) in the San Juan Basin and 1 mmcf/d in the Raton Basin (Rogers et al., 2007). The New York State Revised Draft Supplemental Generic Environmental Impact Statement (NYS rdSGEIS) for the Marcellus Shale cites industry estimates that a typical well will initially produce 2.8 mmcf/d; the production rate will decrease to 550 mcf/d after 5 years and 225 mcf/d after 10 years, after which it will drop approximately 3 percent a year (NYSDEC, 2011). A study of actual production rates in the Barnett Shale found that the average well produces about 800 mmcf during its lifetime, which averages about 7.5 years (Berman, 2009).

Refracturing is possible once an oil or gas well begins to approach the point where it is no longer cost-effectively producing hydrocarbons. Zoback et al. (2010) maintain that shale gas wells are rarely refractured. Berman (2009), however, claims that wells may be refractured once they are no longer profitable. The NYS rdSGEIS estimates that wells may be refractured after roughly five years of service (NYSDEC, 2011).

Once a well is no longer producing gas economically, it is plugged to prevent possible fluid migration that could contaminate soils or waters. According to API, primary environmental concerns include protecting freshwater aquifers and USDWs as well as isolating downhole formations that contain hydrocarbons (API, 2009a). An improperly closed well may provide a pathway for fluid to flow up the well toward ground or surface waters or down the wellbore, leading to contamination of ground water (API, 2009a). A surface plug is used to prevent surface water from seeping into the wellbore and migrating into ground water resources. API recommends setting cement plugs to isolate hydrocarbon and injection/disposal intervals, as well as setting a plug at the base of the lowermost USDW present in the formation (API, 2009a).

3.5 REGULATORY FRAMEWORK

Hydraulic fracturing for oil and gas production wells is typically addressed by state oil and gas boards or equivalent state natural resource agencies. EPA retains authority to address many issues related to hydraulic fracturing under its environmental statutes. The major statutes include the Clean Air Act; the Resource Conservation and Recovery Act; the Clean Water Act; the Safe Drinking Water Act; the Comprehensive Environmental Response, Compensation and Liability Act; the Toxic Substances Control Act; and the National Environmental Policy Act. EPA does not expect to address the efficacy of the regulatory framework as part of this investigation.

4   THE HYDRAULIC FRACTURING WATER LIFECYCLE

The hydraulic fracturing water lifecycle—from water acquisition to wastewater treatment and disposal—is illustrated in Figure 9. The figure also shows potential issues for drinking water resources associated with each phase. Table 1 summarizes the primary and secondary research questions EPA has identified for each stage of the hydraulic fracturing water lifecycle.

The next chapter outlines the research approach and activities needed to answer these questions.

TABLE 1. RESEARCH QUESTIONS IDENTIFIED TO DETERMINE THE POTENTIAL IMPACTS OF HYDRAULIC FRACTURING ON DRINKING WATER RESOURCES

Water Lifecycle Stage	Fundamental Research Question	Secondary Research Questions
Water Acquisition	What are the potential impacts of large volume water withdrawals from ground and surface waters on drinking water resources?	• How much water is used in hydraulic fracturing operations, and what are the sources of this water? • How might withdrawals affect short- and long-term water availability in an area with hydraulic fracturing activity? • What are the possible impacts of water withdrawals for hydraulic fracturing operations on local water quality?
Chemical Mixing	What are the possible impacts of surface spills on or near well pads of hydraulic fracturing fluids on drinking water resources?	• What is currently known about the frequency, severity, and causes of spills of hydraulic fracturing fluids and additives? • What are the identities and volumes of chemicals used in hydraulic fracturing fluids, and how might this composition vary at a given site and across the country? • What are the chemical, physical, and toxicological properties of hydraulic fracturing chemical additives? • If spills occur, how might hydraulic fracturing chemical additives contaminate drinking water resources?
Well Injection	What are the possible impacts of the injection and fracturing process on drinking water resources?	• How effective are current well construction practices at containing gases and fluids before, during, and after fracturing? • Can subsurface migration of fluids or gases to drinking water resources occur and what local geologic or man-made features may allow this? • How might hydraulic fracturing fluids change the fate and transport of substances in the subsurface through geochemical interactions? • What are the chemical, physical, and toxicological properties of substances in the subsurface that may be released by hydraulic fracturing operations? *Table continued on next page*

Table continued from previous page		
Water Lifecycle Stage	**Fundamental Research Question**	**Secondary Research Questions**
Flowback and Produced Water	What are the possible impacts of surface spills on or near well pads of flowback and produced water on drinking water resources?	• What is currently known about the frequency, severity, and causes of spills of flowback and produced water? • What is the composition of hydraulic fracturing wastewaters, and what factors might influence this composition? • What are the chemical, physical, and toxicological properties of hydraulic fracturing wastewater constituents? • If spills occur, how might hydraulic fracturing wastewaters contaminate drinking water resources?
Wastewater Treatment and Waste Disposal	What are the possible impacts of inadequate treatment of hydraulic fracturing wastewaters on drinking water resources?	• What are the common treatment and disposal methods for hydraulic fracturing wastewaters, and where are these methods practiced? • How effective are conventional POTWs and commercial treatment systems in removing organic and inorganic contaminants of concern in hydraulic fracturing wastewaters? • What are the potential impacts from surface water disposal of treated hydraulic fracturing wastewater on drinking water treatment facilities?

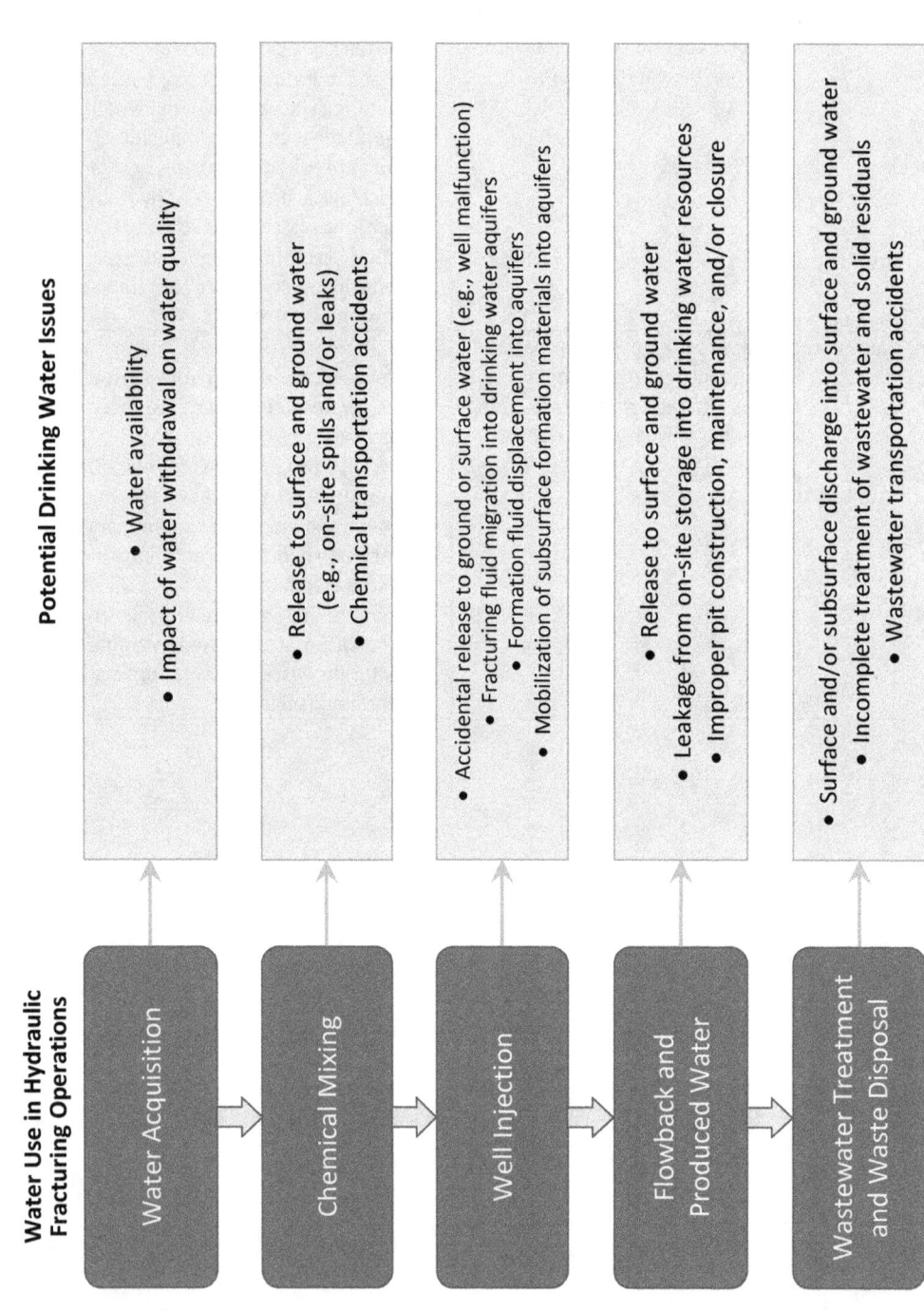

FIGURE 9. WATER USE AND POTENTIAL CONCERNS IN HYDRAULIC FRACTURING OPERATIONS

5   RESEARCH APPROACH

The highly complex nature of the problems to be studied will require a broad range of scientific expertise in environmental and petroleum engineering, ground water hydrology, fate and transport modeling, and toxicology, as well as many other areas. EPA will take a transdisciplinary research approach that integrates various types of expertise from inside and outside EPA. This study uses five main research activities to address the questions identified in Table 1. Table 2 summarizes these activities and their objectives; each activity is then briefly described below with more detailed information available in later chapters.

TABLE 2. RESEARCH ACTIVITIES AND OBJECTIVES

Activity	Objective
Analysis of existing data	Gather and summarize existing data from various sources to provide current information on hydraulic fracturing activities
Case studies	
Retrospective	Perform an analysis of sites with reported contamination to understand the underlying causes and potential impacts to drinking water resources
Prospective	Develop understanding of hydraulic fracturing processes and their potential impacts on drinking water resources
Scenario evaluations	Use computer modeling to assess the potential for hydraulic fracturing to impact drinking water resources based on knowledge gained during existing data analysis and case studies
Laboratory studies	Conduct targeted studies to study the fate and transport of chemical contaminants of concern in the subsurface and during wastewater treatment processes
Toxicological studies	Summarize available toxicological information and, as necessary, conduct screening studies for chemicals associated with hydraulic fracturing operations

5.1   ANALYSIS OF EXISTING DATA

EPA will gather and analyze mapped data on water quality, surface water discharge data, chemical identification data, and site data among others. These data are available from a variety of sources, such as state regulatory agencies, federal agencies, industry, and public sources. Included among these sources are information from the September 2010 letter requesting data from nine hydraulic fracturing service companies and the August 2011 letter requesting data from nine randomly chosen oil and gas well operators. Appendix D contains detailed information regarding these requests.

5.2   CASE STUDIES

Case studies are widely used to conduct in-depth investigations of complex topics and provide a systematic framework for investigating relationships among relevant factors. In addition to reviewing available data associated with the study sites, EPA will conduct environmental field sampling, modeling, and/or parallel laboratory investigations. In conjunction with other elements of the research program, the case studies will help determine whether hydraulic fracturing can impact drinking water resources and, if so, the extent and possible causes of any impacts. Additionally, case studies may provide opportunities to assess the fate and transport of fluids and contaminants in different regions and geologic settings.

Retrospective case studies are focused on investigating reported instances of drinking water resource contamination in areas where hydraulic fracturing events have already occurred. Retrospective case studies will use a deductive logic approach to determine whether or not the reported impacts are due to hydraulic fracturing activity and if so, evaluate potential driving factors for those impacts.

Prospective case studies involve sites where hydraulic fracturing will be implemented after the research begins. These cases allow sampling and characterization of the site prior to, during, and after drilling, water extraction, injection of the fracturing fluid, flowback, and production. At each step in the process, EPA will collect data to characterize both the pre- and post-fracturing conditions at the site. This progressive data collection will allow EPA to evaluate changes in local water availability and quality, as well as other factors, over time to gain a better understanding of the potential impacts of hydraulic fracturing on drinking water resources. Prospective case studies offer the opportunity to sample and analyze flowback and produced water. These studies also provide data to run, evaluate, and improve models of hydraulic fracturing and associated processes, such as fate and transport of chemical contaminants.

5.3   SCENARIO EVALUATIONS

The objective of this approach is to use computer modeling to explore realistic, hypothetical scenarios across the hydraulic fracturing water cycle that may involve adverse impacts to drinking water resources, based primarily on current knowledge and available data. The scenarios will include a reference case involving typical management and engineering practices in representative geologic settings. Typical management and engineering practices will be based on what EPA learns from case studies as well as the minimum requirements imposed by state regulatory agencies. EPA will model surface water in areas to assess impact on water availability and quality where hydraulic fracturing operations withdraw water. EPA will also introduce and model potential modes of failure, both in terms of engineering controls and geologic characteristics, to represent various states of system vulnerability. The scenario evaluations will produce insights into site-specific and regional vulnerabilities.

5.4   LABORATORY STUDIES

Laboratory studies will be used to conduct targeted research needed to better understand the ultimate fate and transport of chemical contaminants of concern. The contaminants of concern may be components of hydraulic fracturing fluids or may be naturally occurring substances released from the subsurface during hydraulic fracturing. Laboratory studies may also be necessary to modify existing analytical methods for case study field monitoring activities. Additionally, laboratory studies will assess the potential for treated flowback or produced water to cause an impact to drinking water resources if released.

5.5   TOXICOLOGICAL STUDIES

Throughout the hydraulic fracturing water lifecycle there are routes through which fracturing fluids and/or naturally occurring substances could be introduced into drinking water resources. To support future risk assessments, EPA will summarize existing data regarding toxicity and potential human health effects associated with these possible drinking water contaminants. Where necessary, EPA may pursue additional toxicological studies to screen and assess the toxicity associated with chemical contaminants of concern.

6   RESEARCH ACTIVITIES ASSOCIATED WITH THE HYDRAULIC FRACTURING WATER LIFECYCLE

This chapter is organized by the hydraulic fracturing water lifecycle depicted in Figure 9 and the associated research questions outlined in Table 1. Each section of this chapter provides relevant background information on the water lifecycle stage and identifies a series of more specific questions that will be researched to answer the fundamental research question. Research activities and expected research outcomes are outlined at the end of the discussion of each stage of the water lifecycle. A summary of the research outlined in this chapter can be found in Appendix A.

6.1   WATER ACQUISITION: WHAT ARE THE POTENTIAL IMPACTS OF LARGE VOLUME WATER WITHDRAWALS FROM GROUND AND SURFACE WATERS ON DRINKING WATER RESOURCES?

6.1.1   BACKGROUND

The amount of water needed in the hydraulic fracturing process depends on the type of formation (coalbed, shale, or tight sands) and the fracturing operations (e.g., well depth and length, fracturing fluid properties, and fracture job design). Water requirements for hydraulic fracturing in coalbed methane range from 50,000 to 350,000 gallons per well (Holditch, 1993; Jeu et al., 1988; Palmer et al., 1991 and 1993). The water usage in shale gas plays is significantly larger: 2 to 4 million gallons of water are typically needed per horizontal well (API, 2010a; GWPC and ALL Consulting, 2009; Satterfield et al., 2008). Table 3 shows how the total volume of water used in fracturing varies depending on the depth and porosity of the shale gas play.

TABLE 3. COMPARISON OF ESTIMATED WATER NEEDS FOR HYDRAULIC FRACTURING OF HORIZONTAL WELLS IN DIFFERENT SHALE PLAYS

Shale Play	Formation Depth (ft)	Porosity (%)	Organic Content (%)	Freshwater Depth (ft)	Fracturing Water (gallons/well)
Barnett	6,500-8,500	4-5	4.5	1,200	2,300,000
Fayetteville	1,000-7,000	2-8	4-10	500	2,900,000
Haynesville	10,500-13,500	8-9	0.5-4	400	2,700,000
Marcellus	4,000-8,500	10	3-12	850	3,800,000

Data are from GWPC and ALL Consulting, 2009.

It was estimated that 35,000 wells were fractured in 2006 alone across the US (Halliburton, 2008). Assuming that the majority of these wells are horizontal wells, the annual national water requirement may range from 70 to 140 billion gallons. This is equivalent to the total amount of water withdrawn from drinking water resources each year in roughly 40 to 80 cities with a population of 50,000 or about one to two cities of 2.5 million people. In the Barnett Shale area, the annual estimates of total water used by gas producers ranged from 2.6 to 5.3 billion gallons per year from 2005 through 2007 (Bene et al., 2007, as cited in Galusky, 2007). During the projected peak shale gas production in 2010, the total water used for gas production in the Barnett Shale was estimated to be 9.5 billion gallons. This represents 1.7 percent of the estimated total freshwater demand by all users within the Barnett Shale area (554 billion gallons) (Galusky, 2007).

To meet these large volume requirements, source water is typically stored in 20,000-gallon portable steel ("frac") tanks located at the well site (GWPC and ALL Consulting, 2009; ICF International, 2009a; Veil, 2007). Source water can also be stored in impoundment pits on site or in a centralized location that services multiple sites. For example, in the Barnett and Fayetteville Shale plays, source water may be stored in large, lined impoundments ranging in capacity from 8 million gallons for 4 to 20 gas wells to 163 million gallons for 1,200 to 2,000 gas wells (Satterfield et al., 2008). The water used to fill tanks or impoundments may come from either ground or surface water, depending on the region in which the fracturing takes place. The transportation of source water to the well site depends on site-specific conditions. In many areas, trucks generally transport the source water to the well site. In the long term, where topography allows, a network of pipelines may be installed to transfer source water between the source and the impoundments or tanks.

Whether the withdrawal of this much water from local surface or ground water sources has a significant impact and the types of possible impacts may vary from one part of the country to another and from one time of the year to another. In arid North Dakota, the projected need of 5.5 billion gallons of water per year to release oil and gas from the Bakken Shale has prompted serious concerns by stakeholders (Kellman and Schneider, 2010). In less arid parts of the country, the impact of water withdrawals may be different. In the Marcellus Shale area, stakeholder concerns have focused on large volume, high rate water withdrawals from small streams in the headwaters of watersheds supplying drinking water (Maclin et al., 2009; Myers, 2009).

One way to offset the large water requirements for hydraulic fracturing is to recycle the flowback produced in the fracturing process. Estimates for the amount of fracturing fluid that is recovered during the first two weeks after a fracture range from 25 to 75 percent of the original fluid injected and depends on several variables, including but not limited to the formation and the specific techniques used (Pickett, 2009; Veil, 2010; Horn, 2009). This water may be treated and reused by adding additional chemicals as well as fresh water to compose a new fracturing solution. There are, however, challenges associated with reusing flowback due to the high concentrations of total dissolved solids (TDS) and other dissolved constituents found in flowback (Bryant et al., 2010). Constituents such as specific cations (e.g., calcium, magnesium, iron, barium, and strontium) and anions (e.g., chloride, bicarbonate, phosphate, and sulfate) can interfere with hydraulic fracturing fluid performance by producing scale or by interfering with chemical additives in the fluids (Godsey, 2011). Recycled water can also become so concentrated with contaminants that it requires either disposal or reuse with considerable dilution. Acid mine drainage, which has a lower TDS concentration, has also been suggested as possible source water for hydraulic fracturing (Vidic, 2010) as well as non-potable ground water, including brackish water, saline, and brine (Godsey, 2011; Hanson, 2011).

6.1.2 HOW MUCH WATER IS USED IN HYDRAULIC FRACTURING OPERATIONS, AND WHAT ARE THE SOURCES OF THIS WATER?

As mentioned in the previous section, source water for hydraulic fracturing operations can come from a variety of sources, including ground water, surface water, and recycled flowback. Water acquisition has not been well characterized, so EPA intends to gain a better understanding of the amounts and sources of water being used for hydraulic fracturing operations.

6.1.2.1 RESEARCH ACTIVITIES – SOURCE WATER

Analysis of existing data. EPA has asked for information on hydraulic fracturing fluid source water resources from nine hydraulic fracturing service companies and nine oil and gas operators (see Appendix D). The data received from the service companies will inform EPA's understanding of the general water quantity and quality requirements for hydraulic fracturing. EPA has asked the nine oil and gas operating companies for information on the total volume, source, and quality of the base fluid[8] needed for hydraulic fracturing at 350 hydraulically fractured oil and gas production wells in the continental US. These data will provide EPA with a nationwide perspective on the volumes and sources of water used for hydraulic fracturing operations, including information on ground and surface water withdrawals as well as recycling of flowback.

EPA will also study water use for hydraulic fracturing operations in two representative regions of the US: the Susquehanna River Basin and Garfield County, Colorado. The Susquehanna River Basin is in the heart of the Marcellus Shale play and represents a humid climate while Garfield County is located in the Piceance Basin and represents a semi-arid climate. EPA will collect existing data from the Susquehanna River Basin Commission and the Colorado Oil and Gas Conservation Commission to determine the volumes of water used for hydraulic fracturing and, if available, the sources of these waters.

EPA expects the research outlined above to produce the following:

- A list of volume and water quality parameters important for hydraulic fracturing operations.
- Information on source, volume, and quality of water used for hydraulic fracturing operations.
- Location-specific data on water use for hydraulic fracturing.

Prospective case studies. EPA will conduct prospective case studies in DeSoto Parish, Louisiana, and Washington County, Pennsylvania. As part of these studies, EPA will monitor the volumes, sources, and quality of water needed for hydraulic fracturing operations. These two locations are representative of an area where ground water withdrawals have been common (Haynesville Shale in Louisiana), and an area where surface water withdrawals and recycling practices have been used (Marcellus Shale in Pennsylvania).

---

[8] In the case of water-based hydraulic fracturing fluids, water would be the base fluid.

EPA expects the research outlined above to produce the following:

- Location-specific examples of water acquisition, including data on the source, volume, and quality of the water.

6.1.3 HOW MIGHT WATER WITHDRAWALS AFFECT SHORT- AND LONG-TERM WATER AVAILABILITY IN AN AREA WITH HYDRAULIC FRACTURING ACTIVITY?

Large volume water withdrawals for hydraulic fracturing are different from withdrawals for other purposes in that much of the water used for the fracturing process may not be recovered after injection. The impact from large volume water withdrawals varies not only with geographic area, but also with the quantity, quality, and sources of the water used. The removal of large volumes of water could stress drinking water supplies, especially in drier regions where aquifer or surface water recharge is limited. This could lead to lowering of water tables or dewatering of drinking water aquifers, decreased stream flows, and reduced volumes of water in surface water reservoirs. These activities could impact the availability of water for drinking in areas where hydraulic fracturing is occurring. The lowering of water levels in aquifers can necessitate the lowering of pumps or the deepening or replacement of wells, as has been reported near Shreveport, Louisiana, in the area of the Haynesville Shale (Louisiana Office of Conservation, 2011).

As the intensity of hydraulic fracturing activities increases within individual watersheds and geologic basins, it is important to understand the net impacts on water resources and identify opportunities to optimize water management strategies.

6.1.3.1 RESEARCH ACTIVITIES – WATER AVAILABILITY

Analysis of existing data. In cooperation with USACE, USGS, state environmental agencies, state oil and gas associations, river basin commissions, and others, EPA will compile data on water use and the hydrology of the Susquehanna River Basin in the Marcellus Shale and Garfield County, Colorado, in the Piceance Basin. These data will include ground water levels, surface water flows, and water quality as well as data on hydraulic fracturing operations, such as the location of wells and the volume of water used during fracturing. These specific study areas represent both arid and humid areas of the country. These areas were chosen based on the availability of data from the Susquehanna River Basin Commission and the Colorado Oil and Gas Conservation Commission.

EPA will conduct simple water balance and geographic information system (GIS) analysis using the existing data. The data collected will be compiled along with information on hydrological trends over the same period of time. EPA will compare control areas with similar baseline water demands and no oil and gas development to areas with intense hydraulic fracturing activity, isolating and identifying any impacts of hydraulic fracturing on water availability. A critical analysis of trends in water flows and water usage patterns will be conducted in areas where hydraulic fracturing activities are occurring to determine whether water withdrawals alter ground and surface water flows. Data collection will support the assessment of the potential impacts of hydraulic fracturing on water availability at various spatial scales (e.g., site, watershed, basin, and play) and temporal scales (e.g., days, months, and years).

EPA expects the research outlined above to produce the following:

- Maps of recent hydraulic fracturing activity and water usage in a humid region (Susquehanna River Basin) and a semi-arid region (Garfield County, Colorado).
- Information on whether water withdrawals for hydraulic fracturing activities alter ground or surface water flows.
- Assessment of impacts of hydraulic fracturing on water availability at various spatial and temporal scales

Prospective case studies. The prospective case studies will evaluate potential short-term impacts on water availability due to large volume water use for hydraulic fracturing in DeSoto Parish, Louisiana, and Washington County, Pennsylvania. The data collected during these case studies will allow EPA to compare potential differences in effects on local water availability between an area where ground water is typically used (DeSoto Parish) and an area where surface water withdrawals are common (Washington County).

EPA expects the research outlined above to produce the following:

- Identification of short-term impacts on water availability from ground and surface water withdrawals associated with hydraulic fracturing activities.

Scenario evaluation. Scenario evaluations will assess potential long-term quantity impacts as a result of cumulative water withdrawals. The evaluations will focus on hydraulic fracturing operations at various spatial and temporal scales in the Susquehanna River Basin and Garfield County, Colorado, using the existing data described above. The scenarios will include at least two futures: (1) average annual conditions in 10 years based on the full exploitation of oil and natural gas resources; and (2) average annual conditions in 10 years based on sustainable water use in hydraulic fracturing operations. Both scenarios will build on predictions for land use and climate (e.g., drought, average, and wet). EPA will take advantage of the future scenario work constructed for the EPA Region 3 Chesapeake Bay Program[9] and the EPA ORD Future Midwestern Landscape Program.[10] The spatial scales of analysis will reflect both environmental boundaries (e.g., site, watershed, river basin, and geologic play) and political boundaries (e.g., city/municipality, county, state, and EPA Region).

These assessments will consider typical water requirements for hydraulic fracturing activities and will also account for estimated demands for water from other human needs (e.g., drinking water, agriculture, and energy), adjusted for future populations. The sustainability analysis will reflect minimum river flow requirements and aquifer drawdown for drought, average, and wet precipitation years, and will allow a determination of the number of typical hydraulic fracturing operations that could be sustained for the relevant formation (e.g., Marcellus Shale) and future scenario. Appropriate physics-based watershed and ground water models will be used for representation of the water balance and hydrologic cycle, as discussed in Chapter 10.

---

[9] http://www.epa.gov/region3/chesapeake/.
[10] http://www.epa.gov/asmdnerl/EcoExposure/FML.html.

EPA expects the research outlined above to produce the following:

- Identification of long-term water quantity impacts on drinking water resources due to cumulative water withdrawals for hydraulic fracturing.

6.1.4 WHAT ARE THE POSSIBLE IMPACTS OF WATER WITHDRAWALS FOR HYDRAULIC FRACTURING OPERATIONS ON LOCAL WATER QUALITY?

Withdrawals of large volumes of ground water can lower the water levels in aquifers. This can affect the aquifer water quality by exposing naturally occurring minerals to an oxygen-rich environment, potentially causing chemical changes that affect mineral solubility and mobility, leading to salination of the water and other chemical contaminations. Additionally, lowered water tables may stimulate bacterial growth, causing taste and odor problems. Depletion of aquifers can also cause an upwelling of lower quality water and other substances (e.g., methane from shallow deposits) from deeper within an aquifer and could lead to subsidence and/or destabilization of the geology.

Withdrawals of large quantities of water from surface water resources (e.g., streams, lakes, and ponds) can significantly affect the hydrology and hydrodynamics of these resources. Such withdrawals from streams can alter the flow regime by changing their flow depth, velocity, and temperature (Zorn et al., 2008). Additionally, removal of significant volumes of water can reduce the dilution effect and increase the concentration of contaminants in surface water resources (Pennsylvania State University, 2010). Furthermore, it is important to recognize that ground and surface water are hydraulically connected (Winter et al., 1998); any changes in the quantity and quality of the surface water can affect ground water and vice versa.

6.1.4.1 RESEARCH ACTIVITIES – WATER QUALITY

Analysis of existing data. EPA will use the data described in Section 6.1.3.1 to analyze changes in water quality in the Susquehanna River Basin and Garfield County, Colorado, to determine if any changes are due to surface or ground water withdrawals for hydraulic fracturing.

EPA expects the research outlined above to produce the following:

- Maps of hydraulic fracturing activity and water quality for the Susquehanna River Basin and Garfield County, Colorado.
- Information on whether water withdrawals for hydraulic fracturing alter local water quality.

Prospective case studies. These case studies will allow EPA to collect data on the quality of ground and surface waters that may be used for hydraulic fracturing before and after water is removed for hydraulic fracturing purposes. EPA will analyze these data to determine if there are any changes in local water quality and if these changes are a result of water withdrawals associated with hydraulic fracturing.

EPA expects the research outlined above to produce the following:

- Identification of impacts on local water quality from withdrawals for hydraulic fracturing.

6.2   CHEMICAL MIXING: WHAT ARE THE POSSIBLE IMPACTS OF SURFACE SPILLS ON OR NEAR WELL PADS OF HYDRAULIC FRACTURING FLUIDS ON DRINKING WATER RESOURCES?

6.2.1   BACKGROUND

Hydraulic fracturing fluids serve two purposes: to create pressure to propagate fractures and to carry the proppant into the fracture. Chemical additives and proppants are typically used in the fracturing fluid. The types and concentrations of chemical additives and proppants vary depending on the conditions of the specific well being fractured, creating a fracturing fluid tailored to the properties of the formation and the needs of the project. In some cases, reservoir properties are entered into modeling programs that simulate fractures (Castle et al., 2005; Hossain and Rahman, 2008). These simulations may then be used to reverse engineer the requirements for fluid composition, pump rates, and proppant concentrations.

Table 4 lists the volumetric composition of a fluid used in a fracturing operation in the Fayetteville Shale as an example of additive types and concentrations (GWPC and ALL Consulting, 2009; API, 2010b). A list of publicly known chemical additives found in hydraulic fracturing fluids is provided in Appendix E.

In the case outlined in Table 4, the total concentration of chemical additives was 0.49 percent. Table 4 also calculates the volume of each additive based on a total fracturing fluid volume of 3 million gallons, and shows that the total volume of chemical additives is 14,700 gallons. In general, the overall concentration of chemical additives in fracturing fluids used in shale gas plays ranges from 0.5 to 2 percent by volume, with water and proppant making up the remainder (GWPC and ALL Consulting, 2009), indicating that 15,000 to 60,000 gallons of the total fracturing fluid consist of chemical additives (assuming a total fluid volume of 3 million gallons).

The chemical additives are typically stored in tanks on site and blended with water and the proppant prior to injection. Flow, pressure, density, temperature, and viscosity can be measured before and after mixing (Pearson, 1989). High pressure pumps then send the mixture from the blender into the well (Arthur et al., 2008). In some cases, special on-site equipment is used to measure the properties of the mixed chemicals *in situ* to ensure proper quality control (Hall and Larkin, 1989).

6.2.2   WHAT IS CURRENTLY KNOWN ABOUT THE FREQUENCY, SEVERITY, AND CAUSES OF SPILLS OF HYDRAULIC FRACTURING FLUIDS AND ADDITIVES?

Large hydraulic fracturing operations require extensive quantities of supplies, equipment, water, and vehicles, which could create risks of accidental releases, such as spills or leaks. Surface spills or releases can occur as a result of tank ruptures, equipment or surface impoundment failures, overfills, vandalism, accidents, ground fires, or improper operations. Released fluids might flow into a nearby surface water body or infiltrate into the soil and near-surface ground water, potentially reaching drinking water aquifers (NYSDEC, 2011).

TABLE 4. AN EXAMPLE OF THE VOLUMETRIC COMPOSITION OF HYDRAULIC FRACTURING FLUID

Component/ Additive Type	Example Compounds	Purpose	Percent Composition (by Volume)	Volume of Chemical (Gallons)[a]
Water		Deliver proppant	90	2,700,000
Proppant	Silica, quartz sand	Keep fractures open to allow gas flow out	9.51	285,300
Acid	Hydrochloric acid	Dissolve minerals, initiate cracks in the rock	0.123	3,690
Friction reducer	Polyacrylamide, mineral oil	Minimize friction between fluid and the pipe	0.088	2,640
Surfactant	Isopropanol	Increase the viscosity of the fluid	0.085	2,550
Potassium chloride		Create a brine carrier fluid	0.06	1,800
Gelling agent	Guar gum, hydroxyethyl cellulose	Thicken the fluid to suspend the proppant	0.056	1,680
Scale inhibitor	Ethylene glycol	Prevent scale deposits in the pipe	0.043	1,290
pH adjusting agent	Sodium or potassium carbonate	Maintain the effectiveness of other components	0.011	330
Breaker	Ammonium persulfate	Allow delayed breakdown of the gel	0.01	300
Crosslinker	Borate salts	Maintain fluid viscosity as temperature increases	0.007	210
Iron control	Citric acid	Prevent precipitation of metal oxides	0.004	120
Corrosion inhibitor	N,N-dimethyl formamide	Prevent pipe corrosion	0.002	60
Biocide	Glutaraldehyde	Eliminate bacteria	0.001	30

Data are from GWPC and ALL Consulting, 2009, and API, 2010b.

[a] Based on 3 million gallons of fluid used.

Over the past few years there have been numerous media reports of spills of hydraulic fracturing fluids (Lustgarten, 2009; M. Lee, 2011; Williams, 2011). While these media reports highlight specific incidences of surface spills of hydraulic fracturing fluids, the frequency and typical causes of these spills remain unclear. Additionally, these reports tend to highlight severe spills. EPA is interested in learning about the range of volumes and reported impacts associated with surface spills of hydraulic fracturing fluids and additives.

6.2.2.1   RESEARCH ACTIVITIES – SURFACE SPILLS OF HYDRAULIC FRACTURING FLUIDS AND ADDITIVES
Analysis of existing data. EPA will compile and evaluate existing information on the frequency, severity, and causes of spills of hydraulic fracturing fluids and additives. These data will come from a variety of sources, including information provided by nine oil and gas operators. In an August 2011 information request sent to these operators, EPA requested spill incident reports for any fluid spilled at 350 different randomly selected well sites in 13 states across the US. Other sources of data are expected to include

spills reported to the National Response Center, state departments of environmental protection (e.g., Pennsylvania and West Virginia), EPA's Natural Gas Drilling Tipline, and others.

EPA will assess the data provided by these sources to reflect a national perspective of reported surface spills of hydraulic fracturing fluids and additives. The goal of this effort is to provide a representative assessment of the frequency, severity, and causes of surface spills associated with hydraulic fracturing fluids and additives.

EPA expects the research outlined above to produce the following:

- Nationwide data on the frequency, severity, and causes of spills of hydraulic fracturing fluids and additives.

6.2.3 WHAT ARE THE IDENTITIES AND VOLUMES OF CHEMICALS USED IN HYDRAULIC FRACTURING FLUIDS, AND HOW MIGHT THIS COMPOSITION VARY AT A GIVEN SITE AND ACROSS THE COUNTRY?

EPA has compiled a list of chemicals that are publicly known to be used in hydraulic fracturing (Table E1 in Appendix E). The chemicals identified in Table E1, however, does not represent the entire set of chemicals used in hydraulic fracturing activities. EPA also lacks information regarding the frequency, quantity, and concentrations of the chemicals used, which is important when considering the toxic effects of hydraulic fracturing fluid additives. Stakeholder meetings and media reports have emphasized the public's concern regarding the identity and toxicity of chemicals used in hydraulic fracturing. Although there has been a trend in recent years of public disclosure of hydraulic fracturing chemicals, inspection of these databases shows that much information is still deemed to be proprietary and is not made available to the public.

6.2.3.1 RESEARCH ACTIVITIES – HYDRAULIC FRACTURING FLUID COMPOSITION

Analysis of existing data. In September 2010, EPA issued information requests to nine hydraulic fracturing service companies seeking information on the identity and quantity of chemicals used in hydraulic fracturing fluid in the past five years (Appendix D). This information will provide EPA with a better understanding of the common compositions of hydraulic fracturing fluids (i.e., identity of components, concentrations, and frequency of use) and the factors that influence these compositions. By asking for data from the past five years, EPA expects to obtain information on chemicals that have been used recently. Some of these chemicals, however, may no longer be used in hydraulic fracturing operations, but could be present in areas where retrospective case studies will be conducted. Much of the data collected from this request have been claimed as confidential business information (CBI). In accordance with 40 C.F.R. Part 2 Subpart B, EPA will treat it as such until a determination regarding the claims is made.

The list of chemicals from the nine hydraulic fracturing service companies will be compared to the list of publicly known hydraulic fracturing chemical additives to determine the accuracy and completeness of the list of chemicals given in Table E1 in Appendix E. The combined list will provide EPA with an inventory of chemicals used in hydraulic fracturing operations.

EPA expects the research outlined above to produce the following:

- Description of types of hydraulic fracturing fluids and their frequency of use (subject to 40 C.F.R. Part 2 Subpart B regulations).
- A list of chemicals used in hydraulic fracturing fluids, including concentrations (subject to 40 C.F.R. Part 2 Subpart B regulations).
- A list of factors that determine and alter the composition of hydraulic fracturing fluids.

Prospective case studies. These case studies will allow EPA to collect information on chemical products used in hydraulic fracturing fluids. EPA will use these data to illustrate how hydraulic fracturing fluids are used at specific wells in the Haynesville and Marcellus Shale plays.

EPA expects the research outlined above to produce the following:

- Illustrative examples of hydraulic fracturing fluids used in the Haynesville and Marcellus Shale plays.

6.2.4 What are the chemical, physical, and toxicological properties of hydraulic fracturing chemical additives?

Chemical and physical properties of hydraulic fracturing chemical additives can help to identify potential human health exposure pathways by describing the mobility of the chemical additives and possible chemical reactions associated with hydraulic fracturing additives. These properties include, but are not limited to: density, melting point, boiling point, flash point, vapor pressure, diffusion coefficients, partition and distribution coefficients, and solubility.

Chemical characteristics can be used to assess the toxicity of hydraulic fracturing chemical additives. Available information may include structure, water solubility, vapor pressure, partition coefficients, toxicological studies, or other factors. There has been considerable public interest regarding the toxicity of chemicals used in hydraulic fracturing fluids. In response to these concerns, the US House of Representatives Committee on Energy and Commerce launched an investigation to examine the practice of hydraulic fracturing in the US. Through this inquiry, the Committee learned that "between 2005 and 2009, the 14 [leading] oil and gas service companies used more than 2,500 hydraulic fracturing products containing 750 chemicals and other components" (Waxman et al., 2011). This included "29 chemicals that are: (1) known or possible human carcinogens; (2) regulated under the Safe Drinking Water Act for their risks to human health; or (3) listed as hazardous air pollutants under the Clean Air Act" (Waxman et al., 2011).

6.2.4.1 Research Activities – Chemical, Physical, and Toxicological Properties

Analysis of existing data. EPA will combine the chemical data collected from the nine hydraulic fracturing service companies with the public list of chemicals given in Appendix E and other sources that may become available to obtain an inventory of chemicals used in hydraulic fracturing fluids. EPA will then search existing databases to obtain known chemical, physical, and toxicological properties for the chemicals in the inventory. EPA expects to use this list to identify a short list of 10 to 20 chemical indicators to track the fate and transport of hydraulic fracturing fluids through the environment. The criteria for selecting these indicators will include, but are not limited to: (1) the frequency of occurrence in fracturing fluids; (2) the toxicity of the chemical; (3) the expected fate and transport of the chemical (e.g., mobility in the environment); and (4) the availability of detection methods. EPA will also use this chemical list to identify chemicals with little or no toxicological information and may be of high concern for human health impacts. These chemicals of concern will undergo further toxicological assessment

EPA expects the research outlined above to produce the following:

- A list of hydraulic fracturing chemicals with known chemical, physical, and toxicological properties.
- Identification of 10-20 possible indicators to track the fate and transport of hydraulic fracturing fluids based on known chemical, physical, and toxicological properties.
- Identification of hydraulic fracturing chemicals that may be of high concern, but have little or no existing toxicological information.

Toxicological analysis/assessment. EPA will identify any hydraulic fracturing chemical currently undergoing ToxCast Phase II testing to determine if chemical, physical, and toxicological properties are being assessed. In other cases where chemical, physical, and toxicological properties are unknown, EPA will estimate these properties using quantitative structure-activity relationships. From this effort, EPA will identify up to six chemicals used in hydraulic fracturing fluid and without toxicity values to be considered for ToxCast screening and provisional peer-reviewed toxicity value (PPRTV) development. More detailed information on characterization of the toxicity and human health approach is found in Chapter 11.

EPA expects the research outlined above to produce the following:

- Lists of high, low, and unknown priority hydraulic fracturing chemicals based on known or predicted toxicity data.
- Toxicological properties for up to six hydraulic fracturing chemicals that have no existing toxicological information and are of high concern.

Laboratory studies. The list of chemicals derived from the existing data analysis and toxicological studies will inform EPA of high priority chemicals for which existing analytical methods may be inadequate for detection in hydraulic fracturing fluids and/or in drinking water resources. EPA will modify these methods to suit the needs of the research.

EPA expects the research outlined above to produce the following:

- Improved analytical methods for detecting hydraulic fracturing chemicals.

6.2.5 IF SPILLS OCCUR, HOW MIGHT HYDRAULIC FRACTURING CHEMICAL ADDITIVES CONTAMINATE DRINKING WATER RESOURCES?

Once released unintentionally into the environment, chemical additives in hydraulic fracturing fluid may contaminate ground water or surface water resources. The pathway by which chemical additives may migrate to ground and surface water depends on many factors, including site-, chemical-, or fluid-specific factors. Site-specific factors refer to the physical characteristics of the site and the spill. These may include the location of the spill with respect to ground and surface water resources, weather conditions at the time of the spill, and the type of surface the spill occurred on (e.g., soil, sand, or plastic liner). Chemical- or fluid-specific factors include the chemical and physical properties of the chemical additives or fluid (e.g., density, solubility, diffusion, and partition coefficients). These properties govern the mobility of the fluid or specific chemical additives through soil and other media. To understand exposure pathways related to surface spills of hydraulic fracturing fluids, EPA must understand site-, chemical-, or fluid-specific factors that govern surface spills.

6.2.5.1 RESEARCH ACTIVITIES – CONTAMINATION PATHWAYS

Analysis of existing data. Surface spills of chemicals, in general, are not restricted to hydraulic fracturing operations and can occur under a variety of conditions. Because these are common problems, there already exists a body of scientific literature that describes how a chemical solution released on the ground can be transported into the subsurface and/or run off to a surface water body. Using the list of hydraulic fracturing fluid chemical additives generated through the research described in Section 6.2.3.1, EPA will identify available data on the fate and transport of hydraulic fracturing fluid additives. The relevant research will be used to assess known impacts of spills of fracturing fluid components on drinking water resources and to identify knowledge gaps related to surface spills of hydraulic fracturing fluid chemical additives.

EPA expects the research outlined above to produce the following:

- Summary of existing research that describes the fate and transport of hydraulic fracturing chemical additives, similar compounds, or classes of compounds.
- Identification of knowledge gaps for future research, if necessary.

Retrospective case studies. Accidental releases from chemical tanks, supply lines or leaking valves have been reported at some of the candidate case study sites (listed in Appendix F) have reported. EPA has identified two locations for retrospective case studies to consider surface spills of hydraulic fracturing fluids through field investigations and sampling: Dunn County, North Dakota, and Bradford and Susquehanna Counties, Pennsylvania. This research will identify any potential impacts on drinking water resources from surface spills, and if impacts were observed, what factors may have contributed to the contamination.

EPA expects the research outlined above to produce the following:

- Identification of impacts (if any) to drinking water resources from surface spills of hydraulic fracturing fluids.
- Identification of factors that led to impacts (if any) to drinking water resources resulting from accidental release of hydraulic fracturing fluids.

6.3 WELL INJECTION: WHAT ARE THE POSSIBLE IMPACTS OF THE INJECTION AND FRACTURING PROCESS ON DRINKING WATER RESOURCES?

6.3.1 BACKGROUND

In a cased well completion, the production casing is perforated prior to the injection of hydraulic fracturing fluid. The perforations allow the injected fluid to enter, and thus fracture, the target formation. Wells can be fractured in either a single stage or multiple stages, as determined by the total length of the injection zone. In a multi-stage fracture, the fracturing operation typically begins with the stage furthest from the wellhead until the entire length of the fracture zone has been fractured.

The actual fracturing process within each stage consists of a series of injections using different volumes and compositions of fracturing fluids (GWPC and ALL Consulting, 2009). Sometimes a small amount of fluid is pumped into the well before the actual fracturing begins. This "mini-frac" may be used to help determine reservoir properties and to enable better fracture design (API, 2009b). In the first stage of the fracture job, fracturing fluid (typically without proppant) is pumped down the well at high pressures to initiate the fracture. The fracture initiation pressure will depend on the depth and the mechanical properties of the formation. A combination of fracturing fluid and proppant is then pumped in, often in slugs of varying sizes and concentrations. After the combination is pumped, a water flush is used to begin flushing out the fracturing fluid (Arthur et al., 2008).

API recommends that several parameters be continuously monitored during the actual hydraulic fracturing process, including surface injection pressure, slurry rate, proppant concentration, fluid rate, and proppant rate (API, 2009b). Monitoring the surface injection pressure is particularly important for two reasons: (1) it ensures that the pressure exerted on equipment does not exceed the tolerance of the weakest components and (2) unexpected or unusual pressure changes may be indicative of a problem that requires prompt attention (API, 2009b). It is not readily apparent how often API's recommendations are followed.

Hydraulic fracturing models and stimulation bottomhole pressure versus time curves can be analyzed to determine fracture height, average fracture width, and fracture half-length. Models can also be used during the fracturing process to make real-time adjustments to the fracture design (Armstrong et al., 1995). Additionally, microseismic monitors and tiltmeters may be used during fracturing to plot the positions of the fractures (Warpinski et al., 1998 and 2001; Cipolla and Wright, 2000), although this is done primarily when a new area is being developed or new techniques are being used (API, 2009b). Comparison of microseismic data to fracture modeling predictions helps to adjust model inputs and increase the accuracy of height, width, and half-length determinations.

6.3.1.1 NATURALLY OCCURRING SUBSTANCES

Hydraulic fracturing can affect the mobility of naturally occurring substances in the subsurface, particularly in the hydrocarbon-containing formation. These substances, described in Table 5, include formation fluid, gases, trace elements, naturally occurring radioactive material, and organic material. Some of these substances may be liberated from the formation via complex biogeochemical reactions with chemical additives found in hydraulic fracturing fluid (Falk et al., 2006; Long and Angino, 1982).

TABLE 5. EXAMPLES OF NATURALLY OCCURRING SUBSTANCES THAT MAY BE FOUND IN HYDROCARBON-CONTAINING FORMATIONS

Type of Contaminant	Example(s)
Formation fluid	Brine[a] (e.g., sodium chloride)
Gases	Natural gas[b] (e.g., methane, ethane), carbon dioxide, hydrogen sulfide, nitrogen, helium
Trace elements	Mercury, lead, arsenic[c]
Naturally occurring radioactive material	Radium, thorium, uranium[c]
Organic material	Organic acids, polycyclic aromatic hydrocarbons, volatile and semi-volatile organic compounds

[a] Piggot and Elsworth, 1996.

[b] Zoback et al., 2010.

[c] Harper, 2008; Leventhal and Hosterman, 1982; Tuttle et al., 2009; Vejahati et al., 2010.

The ability of these substances to reach to ground or surface waters as a result of hydraulic fracturing activities is a potential concern. For example, if fractures extend beyond the target formation and reach aquifers, or if the casing or cement around a wellbore fails under the pressures exerted during hydraulic fracturing, contaminants could migrate into drinking water supplies. Additionally, these naturally occurring substances may be dissolved into or flushed to the surface with the flowback.

6.3.2   HOW EFFECTIVE ARE CURRENT WELL CONSTRUCTION PRACTICES AT CONTAINING GASES AND FLUIDS BEFORE, DURING, AND AFTER FRACTURING?

A number of reports have indicated that that improper well construction or improperly sealed wells may be able to provide subsurface pathways for ground water pollution by allowing contaminant migration to sources of drinking water (PADEP, 2010b; McMahon et al., 2011; State of Colorado Oil and Gas Conservation Commission, 2009a, 2009b, and 2009c; USEPA, 2010b). EPA will assess to what extent proper well construction and mechanical integrity are important factors in preventing contamination of drinking water resources from hydraulic fracturing activities.

In addition to concerns related to improper well construction and well abandonment processes, there is a need to understand the potential impacts of the repeated fracturing of a well over its lifetime. Hydraulic fracturing can be repeated as necessary to maintain the flow of hydrocarbons to the well. The near- and long-term effects of repeated pressure treatments on well construction components (e.g., casing and cement) are not well understood. While EPA recognizes that fracturing or re-fracturing existing wells should also be considered for potential impacts to drinking water resources, EPA has not been able to identify potential partners for a case study; therefore, this practice is not considered in the current study. The issues of well age, operation, and maintenance are important and warrant more study.

6.3.2.1   RESEARCH ACTIVITIES – WELL MECHANICAL INTEGRITY

Analysis of existing data. As part of the voluntary request for information sent by EPA to nine hydraulic fracturing service companies (see Appendix D), EPA asked for the locations of sites where hydraulic fracturing operations have occurred within the past year. From this list of more than 25,000 hydraulic fracturing sites, EPA statistically selected a random sample of sites and requested the complete well files for 350 sites. Well files generally contain information regarding all activities conducted at the site, including any instances of well failure. EPA will analyze the well files to assess the typical frequency, causes, and severity of well failures.

EPA expects the research outlined above to produce the following:

- Data on the frequency and severity of well failures.
- Identification of contributing factors that may lead to well failures during hydraulic fracturing activities.

Retrospective case studies. While conducting retrospective case studies, EPA will assess the mechanical integrity of existing and historical production wells near the reported area of drinking water contamination. To do this, EPA will review existing well construction and mechanical integrity data and/or collect new data using the tools described in Appendix G. EPA will specifically investigate mechanical integrity issues in Dunn County, North Dakota, and Bradford and Susquehanna Counties, Pennsylvania. By investigating well construction and mechanical integrity at sites with reported drinking water contamination, EPA will work to determine if well failure was responsible for the reported contamination and whether original well integrity tests were effective in identifying problems.

EPA expects the research outlined above to produce the following:

- Identification of impacts (if any) to drinking water resources resulting from well failure or improper well construction.
- Data on the role of mechanical integrity in suspected cases of drinking water contamination due to hydraulic fracturing.

Prospective case studies. EPA will evaluate well construction and mechanical integrity at prospective case study sites by assessing the mechanical integrity of the well pre- and post- fracturing. This assessment will be done by comparing results from available logging tools and pressure tests taken before and after hydraulic fracturing. EPA will also assess the methods and tools used to protect drinking water resources from oil and natural gas resources before and during a hydraulic fracture event.

EPA expects the research outlined above to produce the following:

- Data on the changes (if any) in mechanical integrity due to hydraulic fracturing.
- Identification of methods and tools used to isolate drinking water resources from oil and gas resources before and during hydraulic fracturing.

Scenario evaluation. EPA will use computer modeling to investigate the role of mechanical integrity in creating pathways for contaminant migration to ground and surface water resources. The models will include engineering and geological aspects, which will be informed by existing data. Models of the engineering systems will include the design and geometry of the vertical and horizontal wells in addition to information on the casing and cementing materials. Models of the geology will include the expected geometry of aquifers and aquitards/aquicludes, the permeability of the formations, and the geometry and nature of boundary conditions (e.g., closed and open basins, recharge/discharge).

Once built, the models will be used to explore scenarios in which well integrity is compromised before or during hydraulic fracturing due to inadequate or inappropriate well design and construction. In these cases, the construction of the well is considered inadequate due to improper casing and/or cement or improper well construction. It is suspected that breakdowns in the well casing or cement may provide a high permeability pathway between the well casing and the borehole wall, which may lead to contamination of a drinking water aquifer. It will be informative to assess how different types of well construction and testing practices perform during these model scenarios and whether drinking water resources could be affected.

EPA expects the research outlined above to produce the following:

- Assessment of well failure scenarios during and after well injection that may lead to drinking water contamination.

6.3.3 CAN SUBSURFACE MIGRATION OF FLUIDS OR GASES TO DRINKING WATER RESOURCES OCCUR, AND WHAT LOCAL GEOLOGIC OR MAN-MADE FEATURES MAY ALLOW THIS?

Although hydraulic fracture design and control have been researched extensively, predicted and actual fracture lengths still differ frequently (Daneshy, 2003; Warpinski et al., 1998). Hence, it is difficult to accurately predict and control the location and length of fractures. Due to this uncertainty in fracture location, EPA must consider whether hydraulic fracturing may lead to fractures intersecting local geologic or man-made features, potentially creating subsurface pathways that allow fluids or gases to contaminate drinking water resources.

Local geologic features are considered to be naturally occurring features, including pre-existing faults or fractures that lead to or directly extend into aquifers. If the fractures created during hydraulic fracturing were to extend into pre-existing faults or fractures, there may be an opportunity for hydraulic fracturing fluids, natural gas, and/or naturally occurring substances (Table 5) to contaminate nearby aquifers. Any risk posed to drinking water resources would depend on the distance to those resources and the geochemical and transport processes that occur in the intermediate strata. A common assumption in shale gas formations is that natural barriers in the rock strata that act as seals for the gas in the target formation also act as barriers to the vertical migration of fracturing fluids (GWPC and ALL Consulting, 2009). Additionally, during production the flow direction is toward the wellbore because of a decreasing pressure gradient. It is assumed that due to this gradient, gas would be unlikely to move elsewhere as long as the well is in operation and maintains integrity. However, in contrast to shale gas, coalbed methane reservoirs are mostly shallow and may also be co-located with drinking water resources. In this instance, hydraulic fracturing may be occurring in or near a USDW, raising concerns about the contamination of shallow water supplies with hydraulic fracturing fluids (Pashin, 2007).

In addition to natural faults or fractures, it is important to consider the proximity of man-made penetrations such as drinking water wells, exploratory wells, production wells, abandoned wells (plugged and unplugged), injection wells, and underground mines. If such penetrations intersect the injection zone in the vicinity of a hydraulically fractured well, they may serve as conduits for contaminants to reach ground water resources. Several instances of natural gas migrations have been noted. A 2004 EPA report on coalbed methane indicated that methane migration in the San Juan Basin was mitigated once abandoned and improperly sealed wells were plugged. The same report found that in some cases in Colorado, poorly constructed, sealed, or cemented wells used for a variety of purposes could provide conduits for methane migration into shallow USDWs (USEPA, 2004). More recently, a study in the Marcellus Shale region concluded that methane gas was present in well water in areas near hydraulic fracturing operations, but did not identify the origin of the gas (Osborne et al., 2011). Additional studies indicate that methane migration into shallow aquifers is a common natural phenomenon in this region and occurs in areas with and without hydraulic fracturing operations (NYSDEC, 2011).

6.3.3.1   Research Activities – Local Geologic and Man-Made Features

Analysis of existing data. EPA is collecting information from nine oil and gas well operators regarding operations at specific well sites. This information will be compiled and analyzed to determine whether existing local geologic or man-made features are identified prior to hydraulic fracturing, and if so, what types are of concern.

EPA will also review the well files for data relating to fracture location, length, and height. This includes data gathered to measure the fracture pressure gradients in the production zone; data resulting from fracture modeling, microseismic fracture mapping, and/or tiltmeter analysis; and other relevant data. A critical assessment of the available data will allow EPA to determine if fractures created during hydraulic fracturing were localized to the stimulated zone or possibly intersected pre-existing local geologic or man-made features. EPA expects to be able to provide information on the frequency of migration effects and the severity of impacts to drinking water resources posed by these potential contaminant migration pathways.

EPA expects the research outlined above to produce the following:

- Information on the types of local geologic or man-made features identified prior to hydraulic fracturing.
- Data on whether or not fractures interact with local geologic or man-made features and the frequency of occurrence.

Retrospective case studies. In cases of suspected drinking water contamination, EPA will use geophysical testing, field sample analysis, and modeling to investigate the role of local geologic and/or man-made features in leading to any identified contamination. EPA will also review existing data to determine if the induced fractures were confined to the targeted fracture zone. These investigations will determine the role of pre-existing natural or man-made pathways in providing conduits for the migration of fracturing fluid, natural gas, and/or naturally occurring substances to drinking water resources. In particular, EPA will investigate the reported contamination of a USDW in Las Animas County, Colorado, where hydraulic fracturing took place within the USDW.

EPA expects the research outlined above to produce the following:

- Identification of impacts (if any) to drinking water resources from hydraulic fracturing within a drinking water aquifer.

Prospective case studies. The prospective case studies will give EPA a better understanding of the processes and tools used to determine the location of local geologic and/or man-made features prior to hydraulic fracturing. EPA will also evaluate the impacts of local geologic and/or man-made features on the fate and transport of chemical contaminants to drinking water resources by measuring water quality before, during, and after injection. EPA is exploring the possibility of using chemical tracers to track the fate and transport of injected fracturing fluids. The tracers may be used to determine if fracturing fluid migrates from the targeted formation to an aquifer via existing natural or man-made pathways.

EPA expects the research outlined above to produce the following:

- Identification of methods and tools used to determine existing faults, fractures, and abandoned wells.
- Data on the potential for hydraulic fractures to interact with existing natural features.

Scenario evaluation. The modeling tools described above allow for the exploration of scenarios in which the presence of local geologic and man-made features leads to contamination of ground or surface water resources. EPA will explore three different scenarios:

- Induced fractures reaching compromised abandoned wells that intersect and communicate with ground water aquifers.
- Induced fractures reaching ground or surface water resources or permeable formations that communicate with shallower groundwater-bearing strata.
- Sealed or dormant fractures and faults being activated by hydraulic fracturing operations, creating pathways for upward migration of fluids and gases.

In these studies, the injection pulses will be distinguished by their near-field, short-term impacts (fate and transport of injection fluids) as well as their far-field and long-term impacts (including the displacement of native brines or existing gas pockets). These studies will allow the exploration of the potential impacts of fracturing on drinking water resources with regard to variations in geology and will help to inform the retrospective and prospective case studies.

Data provided by these studies will allow EPA to identify and predict the area of evaluation (AOE) around a hydraulic fracturing site. The AOE includes the subsurface zone that may have the potential to be impacted by hydraulic fracturing activities and is projected as an area at the land surface. Within this area, drinking water resources could be affected by the migration of hydraulic fracturing fluids and liberated gases outside the injection zone, as well as the displacement of native brines within the subsurface. Maps of the AOEs for multiple injection operations can be overlaid on regional maps to evaluate cumulative impacts, and, when compared to regional maps of areas contributing recharge to drinking water wells (source water areas), to evaluate regional vulnerability. The AOE may also be used to support contaminant fate and transport hypothesis testing in retrospective case studies.

EPA expects the research outlined above to produce the following:

- Assessment of key conditions that may affect the interaction of hydraulic fractures with existing man-made and natural features.
- Identification of the area of evaluation for a hydraulically fractured well.

6.3.4   HOW MIGHT HYDRAULIC FRACTURING FLUIDS CHANGE THE FATE AND TRANSPORT OF SUBSTANCES IN THE SUBSURFACE THROUGH GEOCHEMICAL INTERACTIONS?

The injection of hydraulic fracturing fluid chemical additives into targeted geologic formations may alter both the injected chemicals and chemicals naturally present in the subsurface. The chemical identity of the injected chemicals may change because of chemical reactions in the fluid (e.g., the formation and breakdown of gels), reactions with the target formation, or microbe-facilitated transformations. These chemical transformation or degradation products could also pose a risk to human health if they migrate to drinking water resources.

Reactions between hydraulic fracturing fluid chemical additives and the target formation could increase or decrease the mobility of these substances, depending on their properties and the complex interactions of the chemical, physical, and biological processes occurring in the subsurface.

For example, several of the chemicals used in fracturing fluid (e.g., acids and carbonates) are known to mobilize naturally occurring substances out of rocks and soils by changing the pH or reduction-oxidation (redox) conditions in the subsurface. Conversely, a change in the redox conditions in the subsurface may also decrease the mobility of naturally occurring substances (Eby, 2004; Sparks, 1995; Sposito, 1989; Stumm and Morgan, 1996; Walther, 2009).

Along with chemical mechanisms, biological processes can change the mobility of fracturing fluid additives and naturally occurring substances. Many microbes, for example, are known to produce siderophores, which can mobilize metals from the surrounding matrix (Gadd, 2004). Microbes may also reduce the mobility of substances by binding to metals or organic substances, leading to the localized sequestration of fracturing fluid additives or naturally occurring substances (Gadd, 2004; McLean and Beveridge, 2002; Southam, 2000).

6.3.4.1   RESEARCH ACTIVITIES – GEOCHEMICAL INTERACTIONS

Laboratory studies. Using samples obtained from retrospective and prospective case study locations, EPA will conduct limited laboratory studies to assess reactions between hydraulic fracturing fluid chemical additives and various environmental materials (e.g., shale or aquifer material) collected on site. Chemical degradation, biogeochemical reactions, and weathering reactions will be studied by pressurizing subsamples of cores, cuttings, or aquifer material in temperature-controlled reaction vessels. Data will be collected on the chemical composition and minerology of these materials. Subsamples will then be exposed to hydraulic fracturing fluids used at the case study locations using either a batch or continuous flow system to simulate subsurface reactions. After specific exposure conditions, samples will be drawn for chemical, mineralogical, and microbiological characterization. This approach will enable the evaluation of the reaction between hydraulic fracturing fluids and environmental media as well as observe chemicals that may be mobilized from the solid phase due to biogeochemical reactions.

EPA expects the research outlined above to produce the following:

- Data on the chemical composition and mineralogy of environmental media.
- Data on the reactions between hydraulic fracturing fluids and environmental media.
- List of chemicals that may be mobilized during hydraulic fracturing activities.

6.3.5   WHAT ARE THE CHEMICAL, PHYSICAL, AND TOXICOLOGICAL PROPERTIES OF SUBSTANCES IN THE SUBSURFACE THAT MAY BE RELEASED BY HYDRAULIC FRACTURING OPERATIONS?

As discussed above, multiple pathways may exist that must be considered for the potential to allow contaminants to reach drinking water resources. These contaminants may include hydraulic fracturing fluid chemical additives and naturally occurring substances, such as those listed in Table 5. Chemical and physical properties of naturally occurring substances can help to identify potential exposure pathways by describing the mobility of these substances and their possible chemical reactions.

The toxic effects of naturally occurring substances can be assessed using toxicological properties associated with the substances. Table E3 in Appendix E provides examples of naturally occurring substances released during hydraulic fracturing operations that may contaminate drinking water resources. The toxicity of these substances varies considerably. For example, some naturally occurring metals, though they can be essential nutrients, exert various forms of toxicity even at low concentrations. Natural gases can also have adverse consequences stemming from their toxicity as well as their physical characteristics (e.g., some are very explosive).

6.3.5.1   RESEARCH ACTIVITIES – CHEMICAL, PHYSICAL, AND TOXICOLOGICAL PROPERTIES

Analysis of existing data. Table E3 in Appendix E lists naturally occurring substances that have been found to be mobilized by hydraulic fracturing activities. EPA will also evaluate data from the literature, as well as from the laboratory studies described above, on the identity of substances and their degradation products released from the subsurface due to hydraulic fracturing. Using this list, EPA will then search existing databases to obtain known chemical, physical, and toxicological properties for these substances. The list will also be used to identify chemicals for further toxicological analysis and analytical method development.

EPA expects the research outlined above to produce the following:

- List of naturally occurring substances that are known to be mobilized during hydraulic fracturing activities and their associated chemical, physical, and toxicological properties.
- Identification of chemicals that may warrant further toxicological analysis or analytical method development.

Toxicological studies. EPA will identify any potential subsurface chemical currently undergoing ToxCast Phase II testing to determine if chemical, physical, and toxicological properties are being assessed. In other cases where chemical, physical, and toxicological properties are unknown, EPA will estimate these properties using quantitative structure-activity relationships. From this effort, EPA will identify up to six chemicals without toxicity values that may be released from the subsurface during hydraulic fracturing for ToxCast screening and PPRTV development consideration. More detailed information on characterization of the toxicity and human health effects of chemicals of concern is found in Chapter 11.

EPA expects the research outlined above to produce the following:

- Lists of high, low, and unknown priority for naturally occurring substances based on known or predicted toxicity data.
- Toxicological properties for up to six naturally occurring substances that have no existing toxicological information and are of high concern.

Laboratory studies. The list of chemicals derived from the existing data analysis and toxicological studies will inform EPA of high priority chemicals for which existing analytical methods may be inadequate for detection in drinking water resources. EPA will modify these methods to suit the needs of the research.

EPA expects the research outlined above to produce the following:

- Analytical methods for detecting selected naturally occurring substances released by hydraulic fracturing.

6.4 FLOWBACK AND PRODUCED WATER: WHAT ARE THE POSSIBLE IMPACTS OF SURFACE SPILLS ON OR NEAR WELL PADS OF FLOWBACK AND PRODUCED WATER ON DRINKING WATER RESOURCES?

6.4.1 BACKGROUND

After the fracturing event, the pressure is decreased and the direction of fluid flow is reversed, allowing fracturing fluid and naturally occurring substances to flow out of the wellbore to the surface before the well is placed into production. This mixture of fluids is called "flowback," which is a subset of produced water. The definition of flowback is not considered to be standardized. Generally, the flowback period in shale gas reservoirs is several weeks (URS Corporation, 2009), while the flowback period in coalbed methane reservoirs appears to be longer (Rogers et al., 2007).

Estimates of the amount of fracturing fluid recovered as flowback in shale gas operations vary from as low as 25 percent to high as 70 to 75 percent (Pickett, 2009; Veil, 2010; Horn, 2009). Other estimates specifically for the Marcellus Shale project a fracture fluid recovery rate of 10 to 30 percent (Arthur et al., 2008). Less information is available for coalbed methane reservoirs. Palmer et al. (1991) estimated a 61 percent fracturing fluid recovery rate over a 19 day period based on sampling from a single well in the Black Warrior Basin.

The flow rate at which the flowback exits the well can be relatively high (e.g., >100,000 gallons per day) for the first few days. However, this flow diminishes rapidly with time, ultimately dropping to the normal rate of produced water flow from a natural gas well (e.g., 50 gallons per day) (Chesapeake Energy, 2010; Hayes, 2009b). While there is no clear transition between flowback and produced water, produced water is generally considered to be the fluid that exits the well during oil or gas production (API, 2010a; Clark and Veil, 2009). Like flowback, produced water also contains fracturing fluid and naturally occurring materials, including oil and/or gas. Produced water, however, is generated throughout the well's lifetime.

The physical and chemical properties of flowback and produced water vary with fracturing fluid composition, geographic location, geological formation, and time (Veil et al., 2004). In general, analyses of flowback from various reports show that concentrations of TDS can range from approximately 1,500 milligram per liter (mg/L) to more than 300,000 mg/L (Gaudlip and Paugh, 2008; Hayes, 2009a; Horn, 2009; Keister, 2009; Vidic, 2010; Rowan et al., 2011). The Appalachian Basin tends to produce one of the higher TDS concentrations by region in the US, with a mean TDS concentration of 250,000 mg/L (Breit, 2002). It can take several weeks for the flowback to reach these values.

Along with high TDS values, flowback can have high concentrations of several ions (e.g., barium, bromide, calcium, chloride, iron, magnesium, sodium, strontium, bicarbonate), with concentrations of calcium and strontium sometimes reported to be as high as thousands of milligrams per liter (Vidic, 2010). Flowback likely contains radionuclides, with the concentration varying by formation (Zielinski and Budahn, 2007; Zoback et al., 2010; Rowan et al., 2011). Flowback from Marcellus Shale formation operations has been measured at concentrations up to 18,000 picocuries per liter (pCi/L; Rowan et al., 2011) and elsewhere in the US above 10,000 pCi/L (USGS, 1999). Volatile organic compounds (VOCs), including but not limited to benzene, toluene, xylenes, and acetone, have also been detected (URS Corporation, 2009; NYSDEC, 2011). A list of chemicals identified in flowback and produced water is presented in Table E2 in Appendix E. Additionally, flowback has been reported to have pH values ranging from 5 to 8 (Hayes, 2009a). A limited time series monitoring program of post-fracturing flowback fluids in the Marcellus Shale indicated increased concentrations over time of TDS, chloride, barium, and calcium; water hardness; and levels of radioactivity (URS Corporation, 2009; Rowen et al., 2011).

Flowback and produced water from hydraulic fracturing operations are held in storage tanks and waste impoundment pits prior to or during treatment, recycling, and disposal (GWPC, 2009). Impoundments may be temporary (e.g., reserve pits for storage) or long-term (e.g., evaporation pits used for treatment). Requirements for impoundments can vary by location. In areas of New York overlying the Marcellus Shale, regulators are requiring water-tight tanks to hold flowback water (ICF, 2009b; NYSDEC, 2011).

6.4.2 WHAT IS CURRENTLY KNOWN ABOUT THE FREQUENCY, SEVERITY, AND CAUSES OF SPILLS OF FLOWBACK AND PRODUCED WATER?

Surface spills or releases of flowback and produced water (collectively referred to as "hydraulic fracturing wastewaters") can occur as a result of tank ruptures, equipment or surface impoundment failures, overfills, vandalism, accidents, ground fires, or improper operations. Released fluids might flow into a nearby surface water body or infiltrate into the soil and near-surface ground water, potentially reaching drinking water aquifers (NYSDEC, 2011). However, it remains unclear how often spills of this nature occur, how severe these spills are, and what causes them. To better understand potential impacts to drinking water resources from surface spills, EPA is interested in learning about the range of volumes and reported impacts associated with surface spills of hydraulic fracturing wastewaters.

6.4.2.1   RESEARCH ACTIVITIES – SURFACE SPILLS OF FLOWBACK AND PRODUCED WATER

Analysis of existing data. EPA will available existing information on the frequency, severity, and causes of spills of flowback and produced water. These data will come from a variety of sources, including information provided by nine oil and gas operators received in response to EPA's August 2011 information request. In this request, EPA asked for spill incident reports for any fluid spilled at 350 different well sites across the US. Other sources of data are expected to include spills reported to the National Response Center, state departments of environmental protection (e.g., Pennsylvania and West Virginia), EPA's Natural Gas Drilling Tipline, and others.

EPA will assess the data provided by these sources to create a national picture of reported surface spills of flowback and produced water. The goal of this effort is to provide a representative assessment of the frequency, severity, and causes of surface spills associated with flowback and produced water.

EPA expects the research outlined above to produce the following:

- Data on the frequency, severity, and common causes of spills of hydraulic fracturing flowback and produced water.

6.4.3   WHAT IS THE COMPOSITION OF HYDRAULIC FRACTURING WASTEWATERS, AND WHAT FACTORS MIGHT INFLUENCE THIS COMPOSITION?

Flowback and produced water can be composed of injected fracturing fluid, naturally occurring materials already present in the target formation, and any reaction or degradation products formed during the hydraulic fracturing process. Much of the existing data on the composition of flowback and produced water focuses on the detection of ions in addition to pH and TDS measurements, as described above. There has been an increased interest in identifying and quantifying the components of flowback and produced water since the composition of these wastewaters affects the treatment and recycling/disposal of the waste (Blauch, 2011; Hayes, 2011; J. Lee, 2011a). However, less is known about the composition and variability of flowback and produced water with respect to the chemical additives found in hydraulic fracturing fluids, reaction and degradation products, or radioactive materials.

The composition of flowback and produced water has also been shown to vary with location and time. For example, data from the USGS produced water database indicate that the distribution of major ions, pH, and TDS levels is not only variable on a national scale (e.g., between geologic basins), but also on the local scale (e.g., within one basin) (USGS, 2002). Studies have also shown that the composition of flowback changes dramatically over time (Blauch, 2011; Hayes, 2011). A better understanding of the spatial and temporal variability of flowback and produced water could lead to improved predictions of the identity and toxicity of chemical additives and naturally occurring substances in hydraulic fracturing wastewaters.

6.4.3.1   RESEARCH ACTIVITIES – COMPOSITION OF FLOWBACK AND PRODUCED WATER

Analysis of existing data. EPA requested data on the composition of flowback and produced water in the information request sent to nine hydraulic fracturing service companies and nine oil and gas operators (Appendix D). EPA will use these data, and any other suitable data it can locate, to better understand what chemicals are likely to be found in flowback and produced water, the variation in chemical concentrations of those chemicals, and what factors may influence their presence and abundance. In this manner, EPA may be able to identify potential chemicals of concern (e.g., fracturing fluid additives, metals, and radionuclides) in flowback and produced water based on their chemical, physical, and toxicological properties.

EPA expects the research outlined above to produce the following:

- A list of chemicals found in flowback and produced water.
- Information on distribution (range, mean, median) of chemical concentrations.
- Identification of factors that may influence the composition of flowback and produced water.
- Identification of the constituents of concern present in hydraulic fracturing wastewaters.

Prospective case studies. EPA will draw samples of flowback and produced water as part of the full water lifecycle monitoring at prospective case study sites. At these sites, flowback and produced water will be sampled periodically following the completion of the injection of hydraulic fracturing fluids into the formation. Samples will be analyzed for the presence of fracturing fluid chemicals and naturally occurring substances found in formation samples analyzed prior to fracturing. This will allow EPA to study the composition and variability of flowback and produced water over a given period of time at two different locations in the Marcellus Shale and the Haynesville Shale.

EPA expects the research outlined above to produce the following:

- Data on composition, variability, and quantity of flowback and produced water as a function of time.

6.4.4   WHAT ARE THE CHEMICAL, PHYSICAL, AND TOXICOLOGICAL PROPERTIES OF HYDRAULIC FRACTURING WASTEWATER CONSTITUENTS?

Chemical, physical, and toxicological properties can be used to aid identification of potential exposure pathways and chemicals of concern related to hydraulic fracturing wastewaters. For example, chemical and physical properties—such as diffusion coefficients, partition, factors and distribution coefficients—can help EPA understand the mobility of different chemical constituents of flowback and produced water in various environmental media (e.g., soil and water). These and other properties will help EPA determine which chemicals in hydraulic fracturing wastewaters may be more likely to appear in drinking water resources. At the same time, toxicological properties can be used to determine chemical constituents that may be harmful to human health. By identifying those chemicals that have a high mobility and substantial toxicity, EPA can identify a set of chemicals of concern associated with flowback and produced water.

6.4.4.1 RESEARCH ACTIVITIES – CHEMICAL, PHYSICAL, AND TOXICOLOGICAL PROPERTIES

Analysis of existing data. EPA will use the data compiled as described in Sections 6.2.3 and 6.4.4 to create a list of chemicals found in flowback and produced water. As outlined in Section 6.2.4, EPA will then search existing databases to obtain known chemical, physical, and toxicological properties for the chemicals in the inventory. EPA expects to identify a list of 10 to 20 chemicals of concern found in hydraulic fracturing wastewaters. The criteria for selecting these chemicals of concern include, but are not limited to: (1) the frequency of occurrence in hydraulic fracturing wastewater; (2) the toxicity of the chemical; (3) the fate and transport of the chemical (e.g., mobility in the environment); and (4) the availability of detection methods.

EPA expects the research outlined above to produce the following:

- List of flowback and produced water constituents with known chemical, physical, and toxicological properties.
- Identification of constituents that may be of high concern, but have no existing toxicological information.

Toxicological studies. EPA will determine if any identified chemical present in flowback or produced water is currently undergoing ToxCast Phase II testing to determine if chemical, physical, and toxicological properties are being assessed. In other cases where chemical, physical, and toxicological properties are unknown, EPA will estimate these properties using quantitative structure-activity relationships. From this effort, EPA will identify up to six chemicals without toxicity values that may be present in hydraulic fracturing wastewaters for ToxCast screening and PPRTV development consideration. More detailed information on characterization of the toxicity and human health effects of chemicals of concern is found in Chapter 11.

EPA expects the research outlined above to produce the following:

- Lists of high, low, and unknown priority chemicals based on known or predicted toxicity data.
- Toxicological properties for up to six hydraulic fracturing wastewater constituents that have no existing toxicological information and are of high concern.

Laboratory studies. The list of chemicals derived from the existing data analysis and toxicological studies will inform EPA of high priority chemicals for which existing analytical methods may be inadequate for detection in hydraulic fracturing wastewaters. EPA will modify these methods to suit the needs of the research.

EPA expects the research outlined above to produce the following:

- Analytical methods for detecting hydraulic fracturing wastewater constituents.

6.4.5   IF SPILLS OCCUR, HOW MIGHT HYDRAULIC FRACTURING WASTEWATERS CONTAMINATE DRINKING WATER RESOURCES?

There may be opportunities for wastewater contamination of drinking water resources both below and above ground. If the mechanical integrity of the well has been compromised, there is the potential for flowback and produced water traveling up the wellbore to have direct access to local aquifers, leading to the contamination of drinking water resources. Once above ground, flowback and produced water are stored on-site in storage tanks and waste impoundment pits, and then may be transported off-site for treatment and/or disposal. There is a potential for releases, leaks, and/or spills associated with the storage and transportation of flowback and produced water, which could lead to contamination of shallow drinking water aquifers and surface water bodies. Problems with the design, construction, operation, and closure of waste impoundment pits may also provide opportunities for releases, leaks, and/or spills. To understand exposure pathways related to surface spills of hydraulic fracturing wastewaters, EPA must consider both site-specific factors and chemical- or fluid-specific factors that govern surface spills (e.g., chemical and physical properties of the fluid).

6.4.5.1   RESEARCH ACTIVITIES – CONTAMINATION PATHWAYS

Analysis of existing data. This approach used here is similar to that described in Section 6.2.5.1 for surface spills associated with the mixing of hydraulic fracturing fluids. Surface spills of chemicals, in general, can occur under a variety of conditions. There already exists a body of scientific literature that describes how a chemical solution released on the ground can infiltrate the subsurface and/or run off to a surface water body. EPA will use the list of chemicals found in hydraulic fracturing wastewaters generated through the research described in Section 6.4.3.1 to identify individual chemicals and classes of chemicals for review in the existing scientific literature. EPA will then identify relevant research on the fate and transport of these chemicals. The research will be summarized to determine the known impacts of spills of fracturing fluid wastewaters on drinking water resources, and to identify existing knowledge gaps related to surface spills of flowback and produced water.

EPA expects the research outlined above to produce the following:

- Summary of existing research that describes the fate and transport of chemicals in hydraulic fracturing wastewaters of similar compounds.
- Identification of knowledge gaps for future research, if necessary.

Retrospective case studies. Accidental releases from wastewater pits and tanks, supply lines, or leaking valves have been reported at some of the candidate case study sites (listed in Appendix F). EPA has identified three retrospective case study locations to investigate surface spills of hydraulic fracturing wastewaters: Wise and Denton Counties, Texas; Bradford and Susquehanna Counties, Pennsylvania; and Washington County, Pennsylvania. The studies will provide an opportunity to identify any impacts to drinking water resources from surface spills. If impacts are found to have occurred, EPA will determine the factors that were responsible for the contamination.

EPA expects the research outlined above to produce the following:

- Identification of impacts (if any) to drinking water resources from surface spills of hydraulic fracturing wastewater.
- Identification of factors that led to impacts (if any) to drinking water resources resulting from the accidental release of hydraulic fracturing wastewaters.

6.5 WASTEWATER TREATMENT AND WASTE DISPOSAL: WHAT ARE THE POSSIBLE IMPACTS OF INADEQUATE TREATMENT OF HYDRAULIC FRACTURING WASTEWATERS ON DRINKING WATER RESOURCES?

6.5.1 BACKGROUND

Wastewaters associated with hydraulic fracturing can be managed through disposal or treatment, followed by discharge to surface water bodies or reuse. Regulations and practices for management and disposal of hydraulic fracturing wastes vary by region and state, and are influenced by local and regional infrastructure development as well as geology, climate, and formation composition. Underground injection is the primary method for disposal in all major gas shale plays, except the Marcellus Shale (Horn, 2009; Veil, 2007 and 2010). Underground injection can be an effective way to manage wastewaters, although insufficient capacity and the costs of trucking wastewater to an injection site can sometimes be problematic (Gaudlip and Paugh, 2008; Veil, 2010).

In shale gas areas near population centers (e.g., the Marcellus Shale), wastewater treatment at publicly owned treatment works (POTWs) or commercial wastewater treatment facilities (CWTs) may be an option for some operations. CWTs may be designed to treat the known constituents in flowback or produced water while POTWs are generally not able to do so effectively. For example, large quantities of sodium and chloride are detrimental to POTW digesters and can result in high TDS concentrations in the effluent (Veil, 2010; West Virginia Water Research Institute, 2010). If the TDS becomes too great in the effluent, it may harm drinking water treatment facilities downstream from POTWs. Additionally, POTWs are not generally equipped to treat fluids that contain radionuclides, which may be released from the formation during hydraulic fracturing. Elevated levels of bromide, a constituent of flowback in many areas, can also create problems for POTWs. Wastewater plants using chlorination as a treatment process will produce more brominated disinfection byproducts (DBPs), which have significant health concerns at high exposure levels. Bromides discharged to drinking water sources may also form DBPs during the treatment process. When POTWs are used, there may be strict limits on the volumes permitted. In Pennsylvania, for example, the disposal of production waters at POTWs is limited to less than 1 percent of the POTW's average daily flow (Pennsylvania Environmental Quality Board, 2009).

As noted earlier, recycling of flowback for use in fracturing other wells is becoming increasingly common and is facilitated by developments in on-site treatment to prepare the flowback for reuse. Researchers at Texas A&M, for example, are developing a mobile treatment system that is being pilot tested in the Barnett Shale (Pickett, 2009). In addition to being used for fracturing other wells, hydraulic fracturing wastewater may be also treated on-site to meet requirements for use in irrigation or for watering livestock (Horn, 2009). Given the logistical and financial benefits to be gained from treatment of flowback water, continued developments in on-site treatment technologies are expected.

6.5.2 WHAT ARE THE COMMON TREATMENT AND DISPOSAL METHODS FOR HYDRAULIC FRACTURING WASTEWATERS, AND WHERE ARE THESE METHODS PRACTICED?

As mentioned earlier, common treatment and disposal methods for hydraulic fracturing wastewaters include underground injection in Class II underground injection control (UIC) wells, treatment followed by surface discharge, and treatment followed by reuse as hydraulic fracturing fluid. Treatment, disposal, and reuse of flowback and produced water from hydraulic fracturing activities are important because of the contaminants present in these waters and their potential for adverse human health impacts. Recent events in West Virginia and Pennsylvania have focused public attention on the treatment and discharge of flowback and produced water to surface waters via POTWs (Puko, 2010; Ward Jr., 2010; Hopey, 2011). The concerns raised by the public have prompted Pennsylvania to request that oil and gas operators not send hydraulic fracturing wastewaters to 15 facilities within the state (Hopey and Hamill, 2011; Legere, 2011). While this issue has received considerable public attention, EPA is aware that many oil and gas operators use UIC wells as their primary disposal option. Treatment and recycling of flowback and produced water are becoming more common in areas where underground injection is not currently feasible.

6.5.2.1 RESEARCH ACTIVITIES – TREATMENT AND DISPOSAL METHODS

Analysis of existing data. As part of the information request to nine oil and gas well operators, EPA asked for information relating to the disposal of wastewater generated at 350 wells across the US. Specifically, EPA asked for the volume and final disposition of flowback and produced water, as well as information relating to recycling of hydraulic fracturing wastewaters (e.g., recycling procedure, volume of fluid recycled, use of recycled fluid, and disposition of any waste generated during recycling). EPA will use the information received to obtain a nationwide perspective of recycling, treatment, and disposal methods currently being used by nine oil and gas operators.

EPA expects the research outlined above to produce the following:

- Nationwide data on recycling, treatment, and disposal methods for hydraulic fracturing wastewaters.

Prospective case studies. While conducting prospective case studies in the Marcellus and Haynesville Shales, EPA will collect information on the types of recycling, treatment, and disposal practices used at the two different locations. These areas are illustrative of a region where UIC wells are a viable disposal option (Haynesville Shale) and where recycling is becoming more common (Marcellus Shale).

EPA expects the research outlined above to produce the following:

- Information on wastewater recycling, treatment, and disposal practices at two specific locations.

6.5.3 HOW EFFECTIVE ARE CONVENTIONAL POTWs AND COMMERCIAL TREATMENT SYSTEMS IN REMOVING ORGANIC AND INORGANIC CONTAMINANTS OF CONCERN IN HYDRAULIC FRACTURING WASTEWATERS?

For toxic constituents that are present in wastewater, their separation and appropriate disposal is the most protective approach for reducing potential adverse impacts on drinking water resources. Much is unknown, however, about the efficacy of current treatment processes for removing certain flowback and produced water constituents, such as fracturing fluid additives and radionuclides. Additionally, the chemical composition and concentration of solid residuals created by wastewater treatment plants that treat hydraulic fracturing wastewater, and their subsequent disposal, warrants more study.

Recycling and reuse of flowback and produced water may not completely alleviate concerns associated with treatment and disposal of hydraulic fracturing wastewaters. While recycling and reuse reduce the immediate need for treatment and disposal—and also reduce water acquisition needs—there will likely be a need to treat and properly dispose of the final concentrated volumes of wastewater from a given area of operation.

6.5.3.1 RESEARCH ACTIVITIES – TREATMENT EFFICACY

Analysis of existing data. EPA will gather existing data on the treatment efficiency and contaminant fate and transport through POTWs and CWTs that have treated hydraulic fracturing wastewaters. Emphasis will be placed on inorganic and organic contaminants, the latter being an area that has the least historical information, and hence the greatest opportunity for advancement in treatment. This information will enable EPA to assess the efficacy of existing treatment options and will also identify areas for further research.

EPA expects the research outlined above to produce the following:

- Collection of analytical data on the efficacy of treatment operations that treat hydraulic fracturing wastewaters.
- Identification of areas for further research.

Laboratory studies. Section 6.4.3.1 describes research on the composition and variability of hydraulic fracturing wastewaters, and on the identification of chemicals of concern in flowback and produced water. This information will be coupled with available data on treatment efficacy to design laboratory studies on the treatability, fate, and transport of chemicals of concern, including partitioning in treatment residues. Studies will be conducted using a pilot-scale wastewater treatment system consisting of a primary clarifier, activated sludge basin, and secondary clarifier. Commercial treatment technologies will also be assessed in the laboratory using actual or synthetic hydraulic fracturing wastewater.

EPA expects the research outlined above to produce the following:

- Data on the fate and transport of hydraulic fracturing water contaminants through wastewater treatment processes, including partitioning in treatment residuals.

Prospective case studies. To the extent possible, EPA will evaluate the efficacy of treatment practices used at the prospective case study locations in Pennsylvania and Louisiana by sampling both pre- and post-treatment wastewaters. It is expected that such studies will include on-site treatment, use of wastewater treatment plants, recycling, and underground injection control wells. In these cases, EPA will identify the fate and transport of hydraulic fracturing wastewater contaminants throughout the treatment and will characterize the contaminants in treatment residuals.

EPA expects the research outlined above to produce the following:

- Data on the efficacy of treatment methods used in two locations.

6.5.4 WHAT ARE THE POTENTIAL IMPACTS FROM SURFACE WATER DISPOSAL OF TREATED HYDRAULIC FRACTURING WASTEWATER ON DRINKING WATER TREATMENT FACILITIES?

Drinking water treatment facilities could be negatively impacted by hydraulic fracturing wastewaters when treatment is followed by surface discharge. For example, there is concern that POTWs may be unable to treat the TDS concentrations potentially found in flowback and produced water, which would lead to high concentrations of both chloride and bromide in the effluent. High TDS levels (>500 mg/L) have been detected in the Monongahela and Youghiogheny Rivers in 2008 and 2010, respectively (J. Lee, 2011b; Ziemkiewicz, 2011). The source of these high concentrations is unknown, however, and they could be due to acid mine drainage treatment plants, active or abandoned coal mines, or shale gas operations. Also, it is unclear how these high TDS concentrations may affect drinking water treatment facilities. It is believed that increased concentrations of chloride and bromide may lead to higher levels of both chlorinated and brominated DBPs at drinking water treatment facilities. The presence of high levels of bromide in waters used by drinking water systems that disinfect through chlorination can lead to higher concentrations of brominated DBPs, which may be of greater concern from a human health perspective than chlorinated DBPs (Plewa and Wagner, 2009). Also, because of their inherent higher molecular weight, brominated DBPs will result in higher concentrations (by weight) than their chlorinated counterparts (e.g., bromoform versus chloroform). This has the potential to cause a drinking water utility to exceed the current DBP regulatory limits.

High chloride and bromide concentrations are not the only factors to be addressed regarding drinking water treatment facilities. Other chemicals, such as naturally occurring radioactive material, may also present a problem to drinking water treatment facilities that are downstream from POTWs or CWTs that ineffectively treat hydraulic fracturing wastewaters. To identify potential impacts to drinking water treatment facilities, it is important to be able to determine concentrations of various classes of chemicals of concern at drinking water intakes.

6.5.4.1 RESEARCH ACTIVITIES – POTENTIAL DRINKING WATER TREATMENT IMPACTS

Laboratory studies. EPA will conduct laboratory studies on the formation of DBPs in hydraulic fracturing-impacted waters (e.g., effluent from a wastewater treatment facility during processing of hydraulic fracturing wastewater), with an emphasis on the formation of brominated DBPs. These studies will explore two sources of brominated DBP formation: hydraulic fracturing chemical additives and high levels of bromide in flowback and produced water. In the first scenario, water samples with known amounts of brominated hydraulic fracturing chemical additives will be equilibrated with chlorine, chloramines, and ozone disinfectants. EPA will then analyze these samples for regulated trihalomethanes (i.e., chloroform, bromoform, bromodichloromethane, and dibromochloromethane), haloacetic acids, and nitrosamines. In the second scenario, EPA will use existing peer-reviewed models to identify problematic concentrations of bromide in source waters.

If actual samples of hydraulic fracturing-impacted source waters can be obtained, EPA will perform laboratory studies to establish baseline parameters for the sample (e.g., existing bromide concentration, total organic concentrations, and pH). The samples will then be subjected to chlorination, chloramination, and ozonation and analyzed for brominated DBPs.

If possible, EPA will identify POTWs or CWTs that are currently treating and discharging hydraulic fracturing wastewaters to surface waters. EPA will then collect discharge and stream samples during times when these treatment facilities are and are not processing hydraulic fracturing wastewaters. This will improve EPA's understanding of how contaminants in the treated effluent change when treated hydraulic fracturing wastewaters are discharged to surface water. EPA will also assess how other sources of contamination (e.g., acid mine drainage) alter contaminant concentrations in the effluent. The goal of this effort is to identify when hydraulic fracturing wastewaters are the cause of high levels of TDS or other contaminants at drinking water treatment facilities.

EPA expects the research outlined above to produce the following:

- Data on the formation of brominated DBPs from chlorination, chloramination, and ozonation treatments of water receiving treated effluent from hydraulic fracturing wastewater treatment.
- Data on the inorganic species in hydraulic fracturing wastewater and other discharge sources that contribute similar species.
- Contribution of hydraulic fracturing wastewater to stream/river contamination.

Scenario evaluations. Scenario evaluations will be used to identify potential impacts to drinking water treatment facilities from surface discharge of treated hydraulic fracturing wastewaters. To accomplish this, EPA will first construct a simplified model of an idealized river section with generalized wastewater treatment discharges and drinking water intakes. To the extent possible, the characteristics of the discharges will be generated based on actual representative information. This model will be able to generate a general guide to releases of treated hydraulic fracturing wastewaters that allows exploration of a range of parameters that may affect drinking water treatment intakes (e.g., discharge rates and concentrations, river flow rates, and distances).

In a second step, EPA will create a watershed-specific scenario that will include the location of specific wastewater and drinking water treatment facilities. Likely candidates for this more detailed scenario include the Monongahela, Allegheny, or Susquehanna River networks. The final choice will be based on the availability of data on several parameters, including the geometry of the river network and flows, and hydraulic fracturing wastewater discharges. The primary result will be an assessment of the potential impacts from disposal practices on specific watersheds. Secondarily, the results of the watershed-specific scenario will be compared to the simplified scenario to determine the ability of the simplified model to capture specific watershed characteristics. Taken together, the two parts of this work will allow EPA to assess the potential impacts of chemicals of concern in flowback and produced water at drinking water treatment intakes.

EPA expects the research outlined above to produce the following:

- Identification of parameters that generate or mitigate drinking water exposure.
- Data on potential impacts in the Monongahela, Allegheny, or Susquehanna River networks.

7 ENVIRONMENTAL JUSTICE ASSESSMENT

Environmental justice is the fair treatment and meaningful involvement of all people regardless of race, color, national origin, or income with respect to the development, implementation, and enforcement of environmental laws, regulations, and policies. Achieving environmental justice is an Agency-wide priority (USEPA, 2010d) and is therefore considered in this study plan.

Stakeholders have raised concerns about the environmental justice implications of gas drilling operations. It has been suggested that people with a lower socioeconomic status may be more likely to consent to drilling arrangements, due to the greater economic need of these individuals, or their more limited ability or willingness to engage with policymakers and agencies. Additionally, since drilling agreements are between landowners and well operators, tenants and neighbors may have little or no input in the decision-making process.

In response to these concerns, EPA has included in the study plan a screening analysis of whether hydraulic fracturing activities may be disproportionately occurring in communities with environmental justice concerns. An initial screening assessment will be conducted to answer the following fundamental research question:

- Does hydraulic fracturing disproportionately occur in or near communities with environmental justice concerns?

Consistent with the framework of the study plan, the environmental justice assessment will focus on the spatial locations of the activities associated with the five stages of the water lifecycle (Figure 1). Each stage of the water lifecycle can be categorized as either occurring onsite (chemical mixing, well injection, and flowback and produced water) or offsite (water acquisition and wastewater treatment/disposal). Because water acquisition, onsite activities and wastewater treatment/disposal generally occur in different locations, EPA has identified three secondary research questions:

- Are large volumes of water for hydraulic fracturing being disproportionately withdrawn from drinking water resources that serve communities with environmental justice concerns?
- Are hydraulically fractured oil and gas wells disproportionately located near communities with environmental justice concerns?

- Is wastewater from hydraulic fracturing operations being disproportionately treated or disposed of (via POTWs or commercial treatment systems) in or near communities with environmental justice concerns?

The following sections outline the research activities associated with each of these secondary research questions.

7.1.1   ARE LARGE VOLUMES OF WATER FOR HYDRAULIC FRACTURING BEING DISPROPORTIONATELY WITHDRAWN FROM DRINKING WATER RESOURCES THAT SERVE COMMUNITIES WITH ENVIRONMENTAL JUSTICE CONCERNS?

7.1.1.1   RESEARCH ACTIVITIES – WATER ACQUISITION LOCATIONS

Analysis of existing data. To the extent data are available, EPA will identify locations where large volume water withdrawals are occurring to support hydraulic fracturing activities. These data will be compared to demographic information from the US Census Bureau on race/ethnicity, income, and age, and then GIS mapping will be used to obtain a visual representation of the data. This will allow EPA to screen for locations where large volume water withdrawals may be disproportionately co-located in or near communities with environmental justice concerns. Locations for further study may be identified, depending on the results of this study.

EPA expects the research outlined above to produce the following:

- Maps showing locations of source water withdrawals for hydraulic fracturing and demographic data.
- Identification of areas where there may be a disproportionate co-localization of hydraulic fracturing water withdrawals and communities with environmental justice concerns.

Prospective case studies. Using data from the US Census Bureau, EPA will also evaluate the demographic profile of communities that may be served by water resources used for hydraulic fracturing of the prospective case study sites.

EPA expects the research outlined above to produce the following:

- Information on the demographic characteristics of communities in or near the two case study sites where hydraulic fracturing water withdrawals occur.

7.1.2   ARE HYDRAULICALLY FRACTURED OIL AND GAS WELLS DISPROPORTIONATELY LOCATED NEAR COMMUNITIES WITH ENVIRONMENTAL JUSTICE CONCERNS?

7.1.2.1   RESEARCH ACTIVITIES – WELL LOCATIONS

Analysis of existing data. As a part of the information request sent by EPA to nine hydraulic fracturing companies (see Appendix C), EPA asked for the locations of sites where hydraulic fracturing operations occurred between 2009 and 2010. EPA will compare these data to demographic information from the US Census Bureau on race/ethnicity, income, and age, and use GIS mapping to visualize the data. An assessment of these maps will allow EPA to screen for locations where hydraulic fracturing may be disproportionately co-located with communities that have environmental justice concerns. Depending upon the outcome of this analysis, locations for further study may be identified.

EPA expects the research outlined above to produce the following:

- Maps showing locations of hydraulically fractured wells (subject to CBI rules) and demographic data.
- Identification of areas where there may be a disproportionate co-localization of hydraulic fracturing well sites and communities with environmental justice concerns.

Retrospective and prospective case studies. EPA will evaluate the demographic profiles of communities near prospective case study sites and communities potentially affected by reported contamination on retrospective case study sites. An analysis of these data will provide EPA with information on the specific communities located at case study locations.

EPA expects the research outlined above to produce the following:

- Information on the demographic characteristics of the communities where hydraulic fracturing case studies were conducted.

7.1.3 IS WASTEWATER FROM HYDRAULIC FRACTURING OPERATIONS BEING DISPROPORTIONATELY TREATED OR DISPOSED OF (VIA POTWS OR COMMERCIAL TREATMENT SYSTEMS) IN OR NEAR COMMUNITIES WITH ENVIRONMENTAL JUSTICE CONCERNS?

7.1.3.1 RESEARCH ACTIVITIES – WASTEWATER TREATMENT/DISPOSAL LOCATIONS

Analysis of existing data. To the extent data are available, EPA will compile a list of wastewater treatment plants accepting wastewater from hydraulic fracturing operations. These data will be compared to demographic information from the US Census Bureau on race/ethnicity, income, and age, and then GIS mapping will be used to visualize the data. This will allow EPA to screen for locations where POTWs and commercial treatment works may be disproportionately co-located near communities with environmental justice concerns, and may identify locations for further study.

EPA expects the research outlined above to produce the following:

- Maps showing locations of hydraulic fracturing wastewater treatment facilities and demographic data.
- Identification of areas where there may be a disproportionate co-localization of hydraulic fracturing wastewater treatment facilities and communities with environmental justice concerns.

Prospective case studies. Using data available from the US Census Bureau, EPA will evaluate the demographic profile of communities near treatment and disposal operations that accept wastewater associated with hydraulic fracturing operations.

EPA expects the research outlined above to produce the following:

- Information on the demographics of communities where treatment and disposal of wastewater from hydraulic fracturing operations at the prospective case study sites has occurred.

8 ANALYSIS OF EXISTING DATA

As outlined in Chapter 6, EPA will evaluate data provided by a variety of stakeholders to answer the research questions posed in Table 1. This chapter describes the types of data EPA will be collecting as well as the approach used for collecting and analyzing these data.

8.1 DATA SOURCES AND COLLECTION

8.1.1 PUBLIC DATA SOURCES

The data described in Chapter 6 will be obtained from a variety of sources. Table 6 provides a selection of public data sources EPA intends to use for the current study. The list in the table is not intended to be comprehensive. EPA will also access data from other sources, including peer-reviewed scientific literature, state and federal reports, and other data sources shared with EPA.

8.1.2 INFORMATION REQUESTS

In addition to publicly available data, EPA has requested information from the oil and gas industry through two separate information requests.[11] The first information request was sent to nine hydraulic fracturing service companies in September 2010, asking for the following information:

- Data on the constituents of hydraulic fracturing fluids—including all chemicals, proppants, and water—used in the last five years.
- All data relating to health and environmental impacts of all constituents listed.
- All standard operating procedures and information on how the composition of hydraulic fracturing fluids may be modified on site.
- All sites where hydraulic fracturing has occurred or will occur within one year of the request date.

The nine companies claimed much of the data they submitted to be CBI. EPA will, in accordance with 40 C.F.R. Part 2 Subpart B, treat these data as such until EPA determines whether or not they are CBI.

A second information request was sent to nine oil and gas well operators in August 2011, asking for the complete well files for 350 oil and gas production wells. These wells were randomly selected from a list of 25,000 oil and gas production wells hydraulically fractured during a one-year period of time. The wells were chosen to illustrate their geographic diversity in the continental US.

---

[11] The complete text of these information requests can be found in Appendix D.

TABLE 6. PUBLIC DATA SOURCES EXPECTED TO BE USED AS PART OF THIS STUDY

Source	Type of Data	Applicable Secondary Research Questions
Susquehanna River Basin Commission	Water use for hydraulic fracturing in the Susquehanna River Basin	• How much water is used in hydraulic fracturing operations, and what are the sources of this water? • What are the possible impacts of water withdrawals for hydraulic fracturing operations on local water quality?
Colorado Oil and Gas Conservation Commission	Water use for hydraulic fracturing in Garfield County, CO	• How much water is used in hydraulic fracturing operations, and what are the sources of this water? • What are the possible impacts of water withdrawals for hydraulic fracturing operations on local water quality?
USGS	Water use in US counties for 1995, 2000, and 2005	• How might withdrawals affect short- and long-term water availability in an area with hydraulic fracturing activity?
State departments of environmental quality or departments of environmental protection	Water quality and quantity Hydraulic fracturing wastewater composition (PA DEP)	• How much water is used in hydraulic fracturing operations, and what are the sources of this water? • What are the possible impacts of water withdrawals for hydraulic fracturing operations on local water quality? • What is the composition of hydraulic fracturing wastewaters, and what factors might influence this composition?
US EPA	Toxicity databases (e.g., ACToR, DSSTox, HERO, ExpoCastDB, IRIS, HPVIS, ToxCastDB, ToxRefDB) Chemical and physical properties databases (e.g., EPI Suite, SPARC)	• What are the chemical, physical, and toxicological properties of hydraulic fracturing chemical additives? • What are the chemical, physical, and toxicological properties of substances in the subsurface that may be released by hydraulic fracturing operations? • What are the chemical, physical, and toxicological properties of hydraulic fracturing wastewater constituents?
National Response Center	Information on spills associated with hydraulic fracturing operations	• What is currently known about the frequency, severity, and causes of spills of hydraulic fracturing fluids and additives? • What is currently known about the frequency, severity, and causes of spills of flowback and produced water?
US Census Bureau	Demographic information from the 2010 Census and the 2005-2009 American Community Survey 5-Year Estimates	• Are large volumes of water for hydraulic fracturing being disproportionately withdrawn from drinking water resources that serve communities with environmental justice concerns? • Are hydraulically fractured oil and gas wells disproportionately located near communities with environmental justice concerns? • Is wastewater from hydraulic fracturing operations being disproportionately treated or disposed of (via POTWs or commercial treatment systems) in or near communities with environmental justice concerns?

8.2 Assuring Data Quality

As indicated in Section 2.6, each research project must have a QAPP, which outlines the necessary QA procedures, quality control activities, and other technical activities that will be implemented for a specific project. Projects using existing data are required to develop data assessment and acceptance criteria for this secondary data. Secondary data will be assessed to determine the adequacy of the data according to acceptance criteria described in the QAPP. All project results will include documentation of data sources and the assumptions and uncertainties inherent within those data.

8.3 Data Analysis

EPA will use the data collected from public sources and information requests to create various outputs, including spreadsheets, GIS maps (if possible), and tables. Data determined to be CBI will be appropriately managed and reported. These outputs will be used to inform answers to the research questions described in Chapter 6 and will also be used to support other research projects, including case studies, additional toxicity assessments, and laboratory studies. A complete summary of research questions and existing data analysis activities can be found in Appendix A.

9 Case Studies

This chapter of the study plan describes the rationale for case study selection as well as the approaches used in both retrospective and prospective case studies.

9.1 Case Study Selection

EPA invited stakeholders nationwide to nominate potential case studies through informational public meetings and by submitting comments electronically or by mail. Appendix F contains a list of the nominated case study sites. Of the 48 nominations, EPA selected seven sites for inclusion in the study: five retrospective sites and two prospective sites. The retrospective case study investigations will focus on locations with reported drinking water contamination where hydraulic fracturing operations have occurred. At the prospective case study sites, EPA will monitor key aspects of the hydraulic fracturing process that cover all five stages of the water cycle.

The final location and number of case studies were chosen based on the types of information a given case study would be able to provide. Table 7 outlines the decision criteria used to identify and prioritize retrospective and prospective case study sites. The retrospective and prospective case study sites were chosen to represent a wide range of conditions that reflect a spectrum of impacts that may result from hydraulic fracturing activities. These case studies are intended to provide enough detail to determine the extent to which conclusions can be generalized at local, regional, and national scales.

TABLE 7. DECISION CRITERIA FOR SELECTING HYDRAULIC FRACTURING SITES FOR CASE STUDIES

Selection Step	Inputs Needed	Decision Criteria
Nomination	• Planned, active, or historical hydraulic fracturing activities • Local drinking water resources • Community at risk • Site location, description, and history • Site attributes (e.g., physical, geology, hydrology) • Operating and monitoring data, including well construction and surface management activities	• Proximity of population and drinking water supplies • Magnitude of activity (e.g., density of wells) • Evidence of impaired water quality (retrospective only) • Health and environmental concerns (retrospective only) • Knowledge gap that could be filled by a case study
Prioritization	• Available data on chemical use, site operations, health, and environmental concerns • Site access for monitoring wells, sampling, and geophysical testing • Potential to collaborate with other groups (e.g., federal, state, or interstate agencies; industry; non-governmental organizations, communities; and citizens)	• Geographic and geologic diversity • Diversity of suspected impacts to drinking water resources • Population at risk • Site status (planned, active, or completed) • Unique geological or hydrological features • Characteristics of water resources (e.g., proximity to site, ground water levels, surface water and ground water interactions, unique attributes) • Multiple nominations from diverse stakeholders • Land use (e.g., urban, suburban, rural, agricultural)

Table 8 lists the retrospective case study locations EPA will investigate as part of this study and highlights the areas to be investigated and the potential outcomes expected for each site. The case study sites listed in Table 8 are illustrative of the types of situations that may be encountered during hydraulic fracturing activities and represent a range of locations. In some of these cases, hydraulic fracturing occurred more than a year ago, while in others, the wells were fractured less than a year ago. EPA expects to be able to coordinate with other federal and state agencies as well as landowners to conduct these studies.

TABLE 8. RETROSPECTIVE CASE STUDY LOCATIONS

Location	Areas to be Investigated	Potential Outcomes	Applicable Secondary Research Questions
Bakken Shale (oil) – Killdeer, Dunn Co., ND	• Production well failure during hydraulic fracturing • Suspected drinking water aquifer contamination • Possible soil contamination	• Identify sources of well failure • Determine if drinking water resources are contaminated and to what extent	• If spills occur, how might hydraulic fracturing chemical additives contaminate drinking water resources? • How effective are current well construction practices at containing gases and fluids before, during, and after fracturing? • Are hydraulically fractured oil and gas wells disproportionately located near communities with environmental justice concerns?
Barnett Shale (gas) – Wise Co., TX	• Spills and runoff leading to suspected drinking water well contamination	• Determine if private water wells and /or drinking water resources are contaminated • Obtain information about mechanisms of transport of contaminants via spills, leaks, and runoff	• If spills occur, how might hydraulic fracturing wastewaters contaminate drinking water resources? • Are hydraulically fractured oil and gas wells disproportionately located near communities with environmental justice concerns?
Marcellus Shale (gas) – Bradford and Susquehanna Cos., PA	• Reported Ground water and drinking water well contamination • Suspected surface water contamination from a spill of fracturing fluids • Reported Methane contamination of multiple drinking water wells	• Determine if drinking water wells and or drinking water resources are contaminated and the source of any contamination • Determine source of methane in private wells • Transferable results due to common types of impacts	• If spills occur, how might hydraulic fracturing chemical additives contaminate drinking water resources? • How effective are current well construction practices at containing gases and fluids before, during, and after fracturing? • Are hydraulically fractured oil and gas wells disproportionately located near communities with environmental justice concerns?

Table continued on next page

61

Table continued from previous page

Location	Areas to be Investigated	Potential Outcomes	Applicable Secondary Research Questions
Marcellus Shale (gas) – Washington Co., PA	• Changes in water quality in drinking water, suspected contamination • Stray gas in wells • Leaky surface pits	• Determine if drinking water resources are impacted and if so, what the sources of any impacts or contamination may be. Identify presence/source of drinking water well contamination • Determine if surface waste storage pits are properly managed to protect surface and ground water	• If spills occur, how might hydraulic fracturing wastewaters contaminate drinking water resources? • Are hydraulically fractured oil and gas wells disproportionately located near communities with environmental justice concerns?
Raton Basin (CBM) – Las Animas and Huerfano Cos., CO	• Potential drinking water well contamination (methane and other contaminants) in an area where hydraulic fracturing is occurring within an aquifer	• Determine source of methane • Determine if drinking water resources are impacted and if so, what the sources of any impacts or contamination may be. Identify presence/source/cause of contamination in drinking water wells	• Can subsurface migration of fluids or gases to drinking water resources occur, and what local geological or man-made features may allow this? • Are hydraulically fractured oil and gas wells disproportionately located near communities with environmental justice concerns?

Prospective case studies are made possible by partnerships with federal and state agencies, landowners, and industry, as highlighted in Appendix A. EPA will conduct prospective case studies in the following areas:

- The Haynesville Shale in DeSoto Parish, Louisiana.
- The Marcellus Shale in Washington County, Pennsylvania.

The prospective case studies will provide information that will help to answer secondary research questions related to all five stages of the hydraulic fracturing water cycle, including:

- How might water withdrawals affect short- and long-term water availability in an area with hydraulic fracturing activity?
- What are the possible impacts of water withdrawals for hydraulic fracturing options on local water quality?
- How effective are current well construction practices at containing gases and fluids before, during, and after fracturing?
- What local geologic or man-made factors may contribute to subsurface migration of fluids or gases to drinking water resources?
- What is the composition of hydraulic fracturing wastewaters, and what factors might influence this composition?
- What are the common treatment and disposal methods for hydraulic fracturing wastewaters, and where are these methods practiced?
- Are large volumes of water being disproportionately withdrawn from drinking water resources that serve communities with environmental justice concerns?
- Are hydraulically fractured oil and gas wells disproportionately located near communities with environmental justice concerns?
- Is wastewater from hydraulic fracturing operations being disproportionately treated or disposed of (via POTWs or commercial treatment systems) in or near communities with environmental justice concerns?

For each case study (retrospective and prospective), EPA will write and approve a QAPP before starting any new data collection, as described in Section 2.6. Upon completion of each case study, a report summarizing key findings will be written, peer reviewed, and published. The data will also be presented in the 2012 and 2014 reports.

The following sections describe the general approaches to be used during the retrospective and prospective case studies. As part of the case studies, EPA will perform extensive sampling of relevant environmental media. Appendix H provides details on field sampling, monitoring, and analytical methods that may be used during both the retrospective and prospective case studies. General information is provided in this study plan, as each case study location is unique.

9.2   RETROSPECTIVE CASE STUDIES

As described briefly in Section 5.2, retrospective case studies are focused on investigating reported instances of drinking water contamination in areas where hydraulic fracturing events have already occurred. Table 8 lists the five locations where EPA will conduct retrospective case studies. Each case study will address one or more stages of the water lifecycle by providing information that will help to answer the research questions posed in Table 1.

While the research questions addressed by each case study vary, there are two goals for all the retrospective case studies: (1) to determine whether or not contamination of drinking water resources has occurred and to what extent; and (2) to assess whether or not the reported contamination is due to hydraulic fracturing activities. These case studies will use available data and may include additional environmental field sampling, modeling, and related laboratory investigations. Additional information on environmental field sampling can be found in Appendix H.

Each retrospective case study will begin by determining the sampling area associated with that specific location. Bounding the scope, vertical, and areal extent of each retrospective case study site will depend on site-specific factors, such as the unique geologic, hydrologic, and geographic characteristics of the site as well as the extent of reported impacts. Where it is obvious that there is only one potential source for a reported impact, the case study site will be fairly contained. Where there are numerous reported impacts potentially involving multiple possible sources, the case study site will be more extensive in all dimensions, making it more challenging to isolate possible sources of drinking water contamination.

The case studies will then be conducted in a tiered fashion to develop integrated data on site history and characteristics, water resources, contaminant migration pathways, and exposure routes. This tiered approach is described in Table 9.

TABLE 9. GENERAL APPROACH FOR CONDUCTING RETROSPECTIVE CASE STUDIES

Tier	Goal	Critical Path
1	Verify potential issue	• Evaluate existing data and information from operators, private citizens, and state agencies • Conduct site visits • Interview stakeholders and interested parties
2	Determine approach for detailed investigations	• Conduct initial sampling: sample wells, taps, surface water, and soils • Identify potential evidence of drinking water contamination • Develop conceptual site model describing possible sources and pathways of the reported contamination • Develop, calibrate, and test fate and transport model(s)
3	Conduct detailed investigations to evaluate potential sources of contamination	• Conduct additional sampling of soils, aquifer, surface water and surface wastewater pits/tanks (if present) • Conduct additional testing: stable isotope analyses, soil gas surveys, geophysical testing, well mechanical integrity testing, and further water testing with new monitoring points • Refine conceptual site model and further test exposure scenarios • Refine fate and transport model(s) based on new information
4	Determine the source(s) of any impacts to drinking water resources	• Develop multiple lines of evidence to determine the source(s) of impacts to drinking water resources • Exclude possible sources and pathways of the reported contamination • Assess uncertainties associated with conclusions regarding the source(s) of impacts

Once the potential issue has been verified in Tier 1, initial sampling activities will be conducted based on the characteristics of the complaints and the nature of the sites. Table 10 lists sample types and testing parameters for initial sampling activities.

TABLE 10. TIER 2 INITIAL TESTING: SAMPLE TYPES AND TESTING PARAMETERS

Sample Type	Testing Parameters
Surface and ground water	• General water quality parameters (e.g., pH, redox potential, dissolved oxygen, TDS) • General water chemistry parameters (e.g., cations and anions, including barium, strontium, chloride, boron) • Metals and metalloids (e.g., arsenic, barium, selenium) • Radionuclides (e.g., radium) • Volatile and semi-volatile organic compounds • Polycyclic aromatic hydrocarbons
Soil	• General water chemistry parameters • Metals • Volatile and semi-volatile organic compounds • Polycyclic aromatic hydrocarbons
Produced water from waste pits or tanks where available	• General water quality parameters • General water chemistry parameters • Metals and metalloids • Radionuclides • Volatile and semi-volatile organic compounds • Polycyclic aromatic hydrocarbons • Fracturing fluid additives/degradates

Results from Tier 1 and initial sampling activities will be used to inform the development of a conceptual site model. The site model will account for the hydrogeology of the location to be studied and be used to determine likely sources and pathways of the reported contamination. The conceptual site model will also be informed by modeling results. These models can help to predict the fate and transport of contaminants, identify appropriate sampling locations, determine possible contamination sources, and understand field measurement uncertainties. The conceptual site model will be continuously updated based on new information, data, and modeling results.

If initial sampling activities indicate potential impacts to drinking water resources, additional testing will be conducted to refine the site conceptual model and further test exposure scenarios (Tier 3). Table 11 describes the additional data to be collected during Tier 3 testing activities.

Results from the tests outlined in Table 11 can be used to further elucidate the sources and pathways of impacts to drinking water resources. These data will be used to support multiple lines of evidence, which will serve to identify the sources of impacts to drinking water resources. EPA expects that it will be necessary to examine multiple lines of evidence in all case studies, since hydraulic fracturing chemicals and contaminants can have other sources or could be naturally present contaminants in shallow drinking water aquifers. The results from all retrospective case study investigations will include a thorough discussion of the uncertainties associated with final conclusions related to the sources and pathways of impacts to drinking water resources.

TABLE 11. TIER 3 ADDITIONAL TESTING: SAMPLE TYPES AND TESTING PARAMETERS

Sample Type / Testing	Testing Parameters
Surface and ground water	• Stable isotopes (e.g., strontium, radium, carbon, oxygen, hydrogen) • Dissolved gases (e.g., methane, ethane, propane, butane) • Fracturing fluid additives
Soil	• Soil gas (e.g., argon, helium, hydrogen, oxygen, nitrogen, carbon dioxide, methane, ethane, propane)
Geophysical testing	• Geologic and hydrogeologic conditions (e.g., faults, fractures, abandoned wells) • Soil and rock properties (e.g., porous media, fractured rock)
Mechanical integrity (review of existing data or testing)	• Casing integrity • Cement integrity
Drill cuttings and core samples	• Metals • Radionuclides • Mineralogical analysis

The data collected during retrospective case studies may be used to assess any risks that may be posed to drinking water resources as a result of hydraulic fracturing activities. Because of this possibility, EPA will develop information on: (1) the toxicity of chemicals associated with hydraulic fracturing; (2) the spatial distribution of chemical concentrations and the locations of drinking water wells; (3) how many people are served by the potentially impacted drinking water resources, including aquifers, wells and or surface waters; and (4) how the chemical concentrations vary over time.

9.3 PROSPECTIVE CASE STUDIES

EPA will conduct two prospective case studies: one in the Marcellus Shale and the other in the Haynesville Shale. In both cases, EPA will have access to the site throughout the process of building and fracturing the well. This access will allow EPA to obtain water quality and other data before pad construction, after pad and well construction, and immediately after fracturing. Additionally, monitoring will continue during a follow-up period of approximately one year after hydraulic fracturing has been completed. Data and methods will be similar to the retrospective case studies, but these studies will allow for baseline water quality sampling, collection of flowback and produced water for analysis, and evaluation of hydraulic fracturing wastewater disposal methods.

The prospective case studies are made possible by partnering with oil and natural gas companies and other stakeholders. Because of the need to enlist the support and collaboration of a wide array of stakeholders in these efforts, case studies of this type will likely be completed 16-24 months from the start dates. However, some preliminary results may be available for the 2012 report.

As in the case of the retrospective studies, each prospective case study will begin by determining the sampling area associated with that specific location. Bounding the scope, vertical, and areal extent of each prospective case study site will depend on site-specific factors, such as the unique geologic, hydrologic, and geographic characteristics of the site. The data collected at prospective case study locations will be placed into a wider regional watershed context. Additionally, the scope of the prospective case studies will encompass all stages of the water lifecycle illustrated in Figure 1.

After the boundaries have been established, the case studies will be conducted in a tiered fashion, as outlined in Table 12.

TABLE 12. GENERAL APPROACH FOR CONDUCTING PROSPECTIVE CASE STUDIES

Tier	Goal	Critical Path
1	Collect existing data	• Gather existing data and information from operators, private citizens, and state agencies • Conduct site visits • Interview stakeholders and interested parties
2	Construct a conceptual site model	• Evaluate existing data • Identify all potential sources and pathways for contamination of drinking water resources • Develop flow system model
3	Conduct field sampling	• Conduct sampling to characterize ground and surface water quality and soil/sediment quality prior to pad construction, following pad and well construction, and immediately after hydraulic fracturing • Collect and analyze time series samples of flowback and produced water • Collect field samples for up to one year after hydraulic fracturing • Calibrate flow system model
4	Determine if there are or are likely to be impacts to drinking water resources	• Analyze data collected during field sampling • Assess uncertainties associated with conclusions regarding the potential for impacts to drinking water resources • Recalibrate flow system model

Results from Tier 1 activities will inform the development of a conceptual site model, which will be used to assess potential pathways for contamination of drinking water resources. This model will help to determine the field sampling activities described in Tier 3. Field sampling will be conducted in a phased approach, as described in Table 13.

The data collected during field sampling activities may also be used to test whether geochemical and hydrologic flow models accurately simulate changes in composition, concentration and or location of hydraulic fracturing fluids over time in different environmental media. These data will be evaluated to determine if there were any impacts to drinking water resources as a result of hydraulic fracturing activities during the limited period of the study. In addition, the data will be evaluated to consider the potential for any future impacts on drinking water resources that could arise after the study period. If impacts are found, EPA will report on the type, cause, and extent of the impacts. The results from all prospective case study investigations will include a discussion of the uncertainties associated with final conclusions related to the potential impacts of hydraulic fracturing on drinking water resources.

TABLE 13. TIER 3 FIELD SAMPLING PHASES

Field Sampling Phases	Critical Path
Baseline characterization of the production well site and areas of concern	• Sample all available existing wells, catalogue depth to drinking water aquifers and their thickness, gather well logs • Sample any adjoining surface water bodies • Sample source water for hydraulic fracturing • Install and sample new monitoring wells • Perform geophysical characterization
Production well construction	• Test mechanical integrity • Resample all wells (new and existing), surface water • Evaluate gas shows from the initiation of surface drilling to the total depth of the well • Assess geophysical logging at the surface portion of the hole
Hydraulic fracturing of the production well	• Sample fracturing fluids • Resample all wells, surface water, and soil gas • Sample flowback • Calibrate and test flow and geochemical models
Gas production	• Resample all wells, surface water, and soil gas • Sample produced water

10 Scenario Evaluations and Modeling

In this study, modeling will integrate a variety of factors to enhance EPA's understanding of potential impacts from hydraulic fracturing on drinking water resources. Modeling will be important in both scenario evaluations and case studies. Scenario evaluations will use existing data to explore potential impacts on drinking water resources in instances where field studies cannot be conducted. In retrospective and prospective case studies, modeling will help identify possible contamination pathways at site-specific locations. The results of modeling activities will provide insight into site-specific and regional vulnerabilities as well as help to identify important factors that affect potential impacts on drinking water resources across all stages of the hydraulic fracturing water lifecycle.

10.1 Scenario Evaluations

Scenario evaluations will be a useful approach for analyzing realistic hypothetical scenarios across the hydraulic fracturing water lifecycle that may result in adverse impacts to drinking water. Specifically, EPA will evaluate scenarios relevant to the water acquisition, well injection, and wastewater treatment and disposal stages of the hydraulic fracturing water lifecycle. In all cases, the scenarios will use information from case studies and minimum state regulatory requirements to define typical management and engineering practices, which will then be used to develop reference cases for the scenarios.

Water acquisition. EPA will evaluate scenarios for two different locations in the US: the Susquehanna River Basin and the Upper Colorado River Basin/Garfield County, Colorado. In these instances, the reference case for the scenarios will be developed using data collected from USGS, the Susquehanna River Basin Commission, and the Colorado Oil and Gas Conservation Commission. The reference case will be associated with the year 2000; this year will be classified as low, median, or high flow based on watershed simulations over the period of 1970-2000.

EPA will then project the water use needs for hydraulic fracturing in the Susquehanna River Basin and Upper Colorado River Basin based on three futures: (1) current business and technology; (2) full natural gas exploitation; and (3) a green technology scenario with sustainable water management practices (e.g., full recycling of produced water), and low population growth. These futures models are described below in more detail. Based on these predictions, EPA will assess the potential impacts of large volume water withdrawals needed for hydraulic fracturing for the period of 2020-2040.

Well injection. EPA will investigate possible mechanisms of well failure and stimulation-induced overburden failure that could lead to upward migration of hydrocarbons, fracturing fluids, and/or brines to ground or surface waters. This will be done through numerical modeling using TOUGH2 with geomechanical enhancements. The scenarios also include multiple injection and pumping wells and the evaluations of diffuse and focused leakage (through fractures and abandoned unplugged wells) within an area of potential influence. The reference cases will be determined from current management and engineering practices as well as representative geologic settings. The failure scenarios are described in greater detail in Section 6.3.2.1.

Wastewater treatment and disposal. EPA will use a staged approach to evaluate the potential for impacts of releases of treated hydraulic fracturing wastewaters to surface waters. The first approach will focus on basic transport processes occurring in rivers and will be based on generalized inputs and receptor locations. This work will use scenarios representing various flow conditions, distances between source and receptor, and available data on possible discharge concentrations. The chemicals of interest are the likely residues in treated wastewater, specifically chloride, bromide and naturally occurring radioactive materials. In the second stage, specific watersheds will be evaluated using the best data available for evaluations. Similar to the first stage, scenarios will be developed to show how various conditions in the actual river networks impact concentrations at drinking water receptors. A comparison of both stages will help show the level of detail necessary for specific watersheds and might lead to revision of the first, or more generic, approach.

10.2 CASE STUDIES

Modeling will be used in conjunction with data from case studies to gain a better understanding of the potential impacts of hydraulic fracturing on drinking water resources. First, models will be developed to simulate the flow and transport of hydraulic fracturing fluids and native fluids in an oil or gas reservoir during the hydraulic fracturing process. These models will use data from case studies—including injection pressures, flow rates, and lithologic properties—to simulate the development of fractures and migration of fracturing fluids in the fracture system induced by the hydraulic fracturing process. The results of the modeling may be used to help predict the possibility of rock formation damage and the spreading area of fracturing fluid. Expected outputs include information on the possibility that hydraulic fracturing-related contaminants will migrate to an aquifer system.

Models can also be developed to simulate flow and transport of the contaminants once migration to an aquifer occurs. This modeling will consider a relatively large-scale ground water aquifer system. The modeling will consider the possible sources of fracturing fluids emerging from the oil or gas reservoir through a damaged formation, geological faults, or an incomplete cementing zone outside the well casing. It will also consider local hydrogeological conditions such as precipitation, water well distribution, aquifer boundaries, and hydraulic linkage with other water bodies. The modeling will simulate ground water flow and transport in the aquifer system, and is expected to output information on contamination occurring near water supply facilities. This modeling may also provide the opportunity to answer questions about potential risks associated with hypothetical scenarios, such as conditions under which an improperly cemented wellbore might release fracturing fluid or native fluids (including native gases).

10.3 MODELING TOOLS

EPA expects that a wide range of modeling tools may be used in this study. It is standard practice to evaluate and model complex environmental systems as separate components, as can be the case with potential impacts to drinking water resources associated with hydraulic fracturing. For example, system components can be classified based on media type, such as water body models, ground water models, watershed models, and waste unit models. Additionally, models can be chosen based on whether a stochastic or deterministic representation is needed, solution types (e.g., analytical, semi-analytical, or numerical), spatial resolution (e.g., grid, raster, or vector), or temporal resolution (e.g., steady-state or time-variant).

The types of models to be used in this study may include:

Hydraulic fracturing models. EPA is considering using MFrac to calculate the development of fracture systems during real-time operations. MFrac is a comprehensive design and evaluation simulator containing a variety of options, including three-dimensional fracture geometry and integrated acid fracturing solutions. EPA may also use MFrac to assess formation damage subject to various engineering operations, lithostratigraphy, and depositional environment of oil and gas deposits.

Multi-phase and multi-component ground water models. Members of the TOUGH family of models developed at Lawrence Berkeley National Laboratory can be used to simulate the flow and transport phenomena in fractured zones, where geothermal and geochemical processes are active, where permeability changes, and where phase-change behavior is important. These codes have been adapted for problems requiring capabilities that will be also needed for hydraulic fracturing simulation: multiphase and multi-component transport, geothermal reservoir simulation, geologic sequestration of carbon, geomechanical modeling of fracture activation and creation, and inverse modeling.

Single-phase and multi-component ground water models. These ground water models include:

- The finite difference solutions, such as the USGS Modular Flow and its associated transport codes, including Modular Transport 3D-Multispecies and the related Reactive Transport 3D,
- The finite element solutions, such as the Finite Element Subsurface Flow Model and other semi-analytical solutions (e.g., GFLOW and TTim).

Various chemical and/or biological reactions can be integrated into the advective ground water flow models to allow the simulation of reaction flow and transport in the aquifer system. For a suitably conceptualized system consisting of single-phase transport of water-soluble chemicals, these models can support hydraulic fracturing assessments.

Watershed models. EPA has experience with the well-established watershed management models Soil Water Assessment Tool (semi-empirical, vector-based, continuous in time) and Hydrologic Simulation Program – FORTRAN (semi-physics-based, vector-based, continuous in time). The watershed models will play an important role in modeling water acquisition and in water quantity analysis.

Waterbody models. The well-established EPA model for representing water quality in rivers and reservoirs is the Water Quality Analysis Simulation Program. Other, simpler approaches include analytical solutions to the transport equation and models such as a river and stream water quality model (QUAL2K; see Chapra, 2008). Based on extensive tracer studies, USGS has developed empirical relationships for travel time and longitudinal dispersion in rivers and streams (Jobson, 1996).

Alternative futures models. Alternative futures analysis has three basic components (Baker et al., 2004): (1) characterize the current and historical landscapes in a geographic area and the trajectory of the landscape to date; (2) develop two or more alternative "visions" or scenarios for the future landscape that reflect varying assumptions about land and water use and the range of stakeholder viewpoints; and (3) evaluate the likely effects of these landscape changes and alternative futures on things people care about (e.g., valued endpoints). EPA has conducted alternative futures analysis for much of the landscape of interest for this project. The Agency has created futures for 20 watersheds[12] across the country, including the Susquehanna River basin, which overlays the Marcellus Shale and the Upper Colorado River Basin, which includes Garfield County, Colorado.

---

[12] http://cfpub.epa.gov/ncea/global/recordisplay.cfm?deid=212763

10.4 Uncertainty in Model Applications

All model parameters are uncertain because of measurement approximation and error, uncharacterized point-to-point variability, reliance on estimates and imprecise scale-up from laboratory measurements. Model outputs are subject to uncertainty, even after model calibration (e.g., Tonkin and Dougherty, 2008; Doherty, 2011). Thus, environmental models do not possess generic validity (Oreskes et al., 1994), and the application is critically dependent on choices of input parameters, which are subject to the uncertainties described above. Further, a recent review by one of the founders of the field of subsurface transport modeling (Leonard F. Konikow) outlines the difficulties with contaminant transport modeling and concludes that "Solute transport models should be viewed more for their value in improving the understanding of site-specific processes, hypothesis testing, feasibility assessments, and evaluating data-collection needs and priorities; less value should be placed on expectations of predictive reliability" (Konikow, 2010). Proper application of models requires proper expectations (i.e., Konikow, 2010) and acknowledgement of uncertainties, which can lead to best scientific credibility for the results (see Oreskes, 2003).

11 Characterization of Toxicity and Human Health Effects

EPA will evaluate all stages of the hydraulic fracturing water lifecycle to assess the potential for fracturing fluids and/or naturally occurring substances to be introduced into drinking water resources. As highlighted throughout Chapter 6, EPA will assess the toxicity and potential human health effects associated with these possible drinking water contaminants. To do this, EPA will first obtain an inventory of the chemicals associated with hydraulic fracturing activities (and their estimated concentrations and frequency of occurrence). This includes chemicals used in hydraulic fracturing fluids, naturally occurring substances that may be released from subsurface formations during the hydraulic fracturing process, and chemicals that are present in hydraulic fracturing wastewaters. EPA will also identify the relevant reaction and degradation products of these substances—which may have different toxicity and human health effects than their parent compounds—in addition to the fate and transport characteristics of the chemicals. The aggregation of these data is described in Chapter 6.

Based on the number of chemicals currently known to be used in hydraulic fracturing operations, EPA anticipates that there could be several hundred chemicals of potential concern for drinking water resources. Therefore, EPA will develop a prioritized list of chemicals and, where estimates of toxicity are not otherwise available, conduct quantitative health assessments or additional testing for certain high-priority chemicals. In the first phase of this work, EPA will conduct an initial screen for known toxicity and human health effects information (including existing toxicity values such as reference doses and cancer slope factors) by searching existing databases.[13] At this stage, chemicals will be grouped into one of three categories: (1) high priority for chemicals that are potentially of concern; (2) low priority for

---

[13] These databases include the Integrated Risk Information System (IRIS), the Provisional Peer Reviewed Toxicity Value (PPRTV) database, the ATSDR Minimal Risk Levels (MRLs), the California EPA Office of Environmental Health Hazard Assessment (OEHHA) Toxicity Criteria Database (TCD). Other Agency databases including the Distributed Structure Searchable Toxicity (DSSTox) database, Aggregated Computational Toxicology Resources (ACToR) database and the Toxicity Reference Database (ToxRefDB) may be used to facilitate data searching activities.

chemicals that are likely to be of little concern; and (3) unknown priority for chemicals with an unknown level of concern. These groupings will be based on known chemical, physical, and toxicological properties; reported occurrence levels; and the potential need for metabolism information.

Chemicals with an unknown level of concern are those for which no toxicity information is available. For these chemicals, a quantitative structure-activity relationships (QSAR) analysis may be conducted to obtain comparative toxicity information. A QSAR analysis uses mathematical models to predict measures of toxicity from physical/chemical characteristics of the structure of the chemicals. This approach may provide information to assist EPA in designating these chemicals as either high or low priority.

The second phase of this work will focus on additional testing and/or assessment of chemicals with an unknown level of concern. These chemicals may be subjected to a battery of tests used in the ToxCast program, a high-throughput screening tool that can identify toxic responses (Judson et al., 2010a and 2010b; Reif et al., 2010). The quantitative nature of these *in vitro* assays provides information on concentration-response relationships that, tied to known modes of action, can be useful in assessing the level of potential toxicity. EPA will identify a small set of these chemicals with unknown toxicity values and develop ToxCast bioactivity profiles and hazard predictions for these chemicals.

EPA will use these ToxCast profiles, in addition to existing information, to develop chemical-specific Provisional Peer Reviewed Toxicity Values (PPRTVs) for up to six of the highest-priority chemicals that have no existing toxicity values. PPRTVs summarize the available scientific information about the adverse effects of a chemical and the quality of the evidence, and ultimately derive toxicity values, such as provisional reference doses and cancer slope factors, that can be used in conjunction with exposure and other information to develop a risk assessment. Although using ToxCast is suitable for many of the chemicals used in hydraulic fracturing, the program has excluded any chemicals that are volatile enough to invalidate their assays.

In addition to single chemical assessments, further information may be obtained for mixtures of chemicals based on which components occur most frequently together and their relevant proportions as identified from exposure information. It may be possible to test actual hydraulic fracturing fluids or wastewater samples. EPA will assess the feasibility of this research and pursue testing if possible.

EPA anticipates that the initial database search and ranking of high, low, and unknown priority chemicals will be completed for the 2012 interim report. Additional work using QSAR analysis and high-throughput screening tools is expected to be available in the 2014 report. The development of chemical-specific PPRTVs for high-priority chemicals is also expected to be available in 2014.

Information developed from this effort to characterize the toxicity and health effects of chemicals will be an important component of future efforts to understand the overall potential risk posed by hydraulic fracturing chemicals that may be present in drinking water resources. When combined with exposure and other relevant data, this information will help EPA characterize the potential public health impacts of hydraulic fracturing on drinking water resources.

12 SUMMARY

The objective of this study is to assess the potential impacts of hydraulic fracturing on drinking water resources and to identify the driving factors that affect the severity and frequency of any impacts. The research outlined in this document addresses all stages of the hydraulic fracturing water lifecycle shown in Figure 1 and the research questions posed in Table 1. In completing this research, EPA will use available data, supplemented with original research (e.g. case studies, generalized scenario evaluations and modeling) where needed. As the research progresses, EPA may learn certain information that suggests that modifying the initial approach or conducting additional research within the overall scope of the study plan is prudent in order to better answer the research questions. In that case, EPA may modify the current research plan. Figures 10 and 11 summarize the research activities for the study plan and reports anticipated timelines for research results. All data, whether generated by the EPA or not, will undergo a comprehensive quality assurance.

Water Acquisition → Chemical Mixing → Well Injection

Retrospective Case Studies

Investigate the location, cause, and impact of surface spills/accidental releases of hydraulic fracturing fluids

Investigate the role of mechanical integrity, well construction, and geologic/man-made features in suspected cases of drinking water contamination

Prospective Case Studies

Identify chemical products used in hydraulic fracturing fluids at case study locations

Identify methods and tools used to protect drinking water from oil and gas resources before and after hydraulic fracturing

Assess potential for hydraulic fractures to interfere with existing geologic features

Analysis of Existing Data

Compile information on the frequency, severity, and causes of spills of hydraulic fracturing fluids

Analyze data obtained from 350 well files

Compile data on the composition of hydraulic fracturing fluids

Identify possible chemical indicators and existing analytical methods

Review existing scientific literature on surface chemical spills

Document the source, quality, and quantity of water used for hydraulic fracturing

Evaluate impacts on local water quality and availability from water withdrawals

Compile and analyze existing data on source water volume and quality requirements

Collect data on water use, hydrology, and hydraulic fracturing activities in an arid and humid region

Identify known chemical, physical, and toxicological properties of chemicals found in hydraulic fracturing fluids and naturally occurring chemicals released during hydraulic fracturing

Results expected for 2012 report
Results expected for 2014 report

FIGURE 10A. SUMMARY OF RESEARCH PROJECTS PROPOSED FOR THE FIRST THREE STAGES OF THE HYDRAULIC FRACTURING WATER LIFECYCLE

75

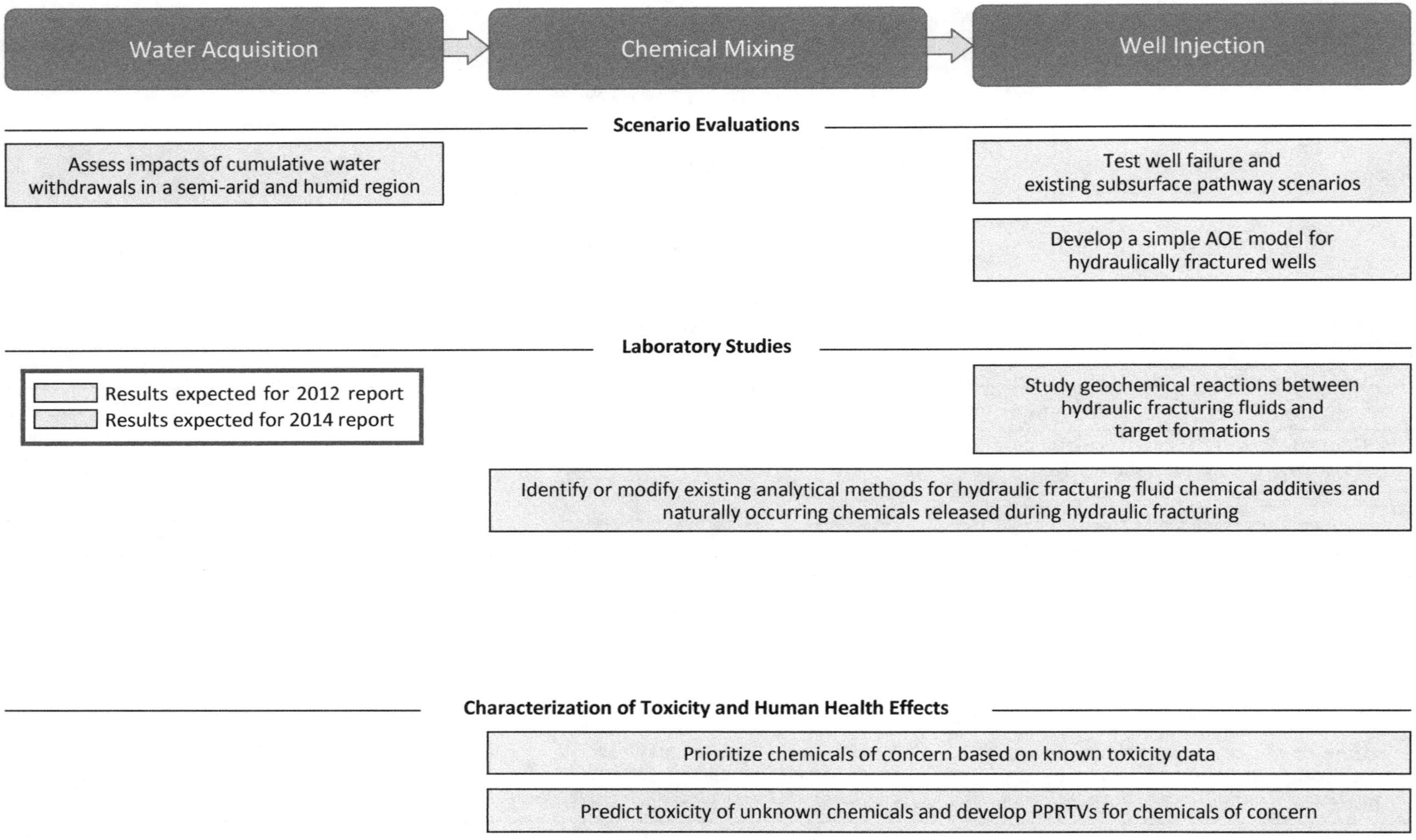

FIGURE 10B. SUMMARY OF RESEARCH PROJECTS PROPOSED FOR THE FIRST THREE STAGES OF THE HYDRAULIC FRACTURING WATER LIFECYCLE

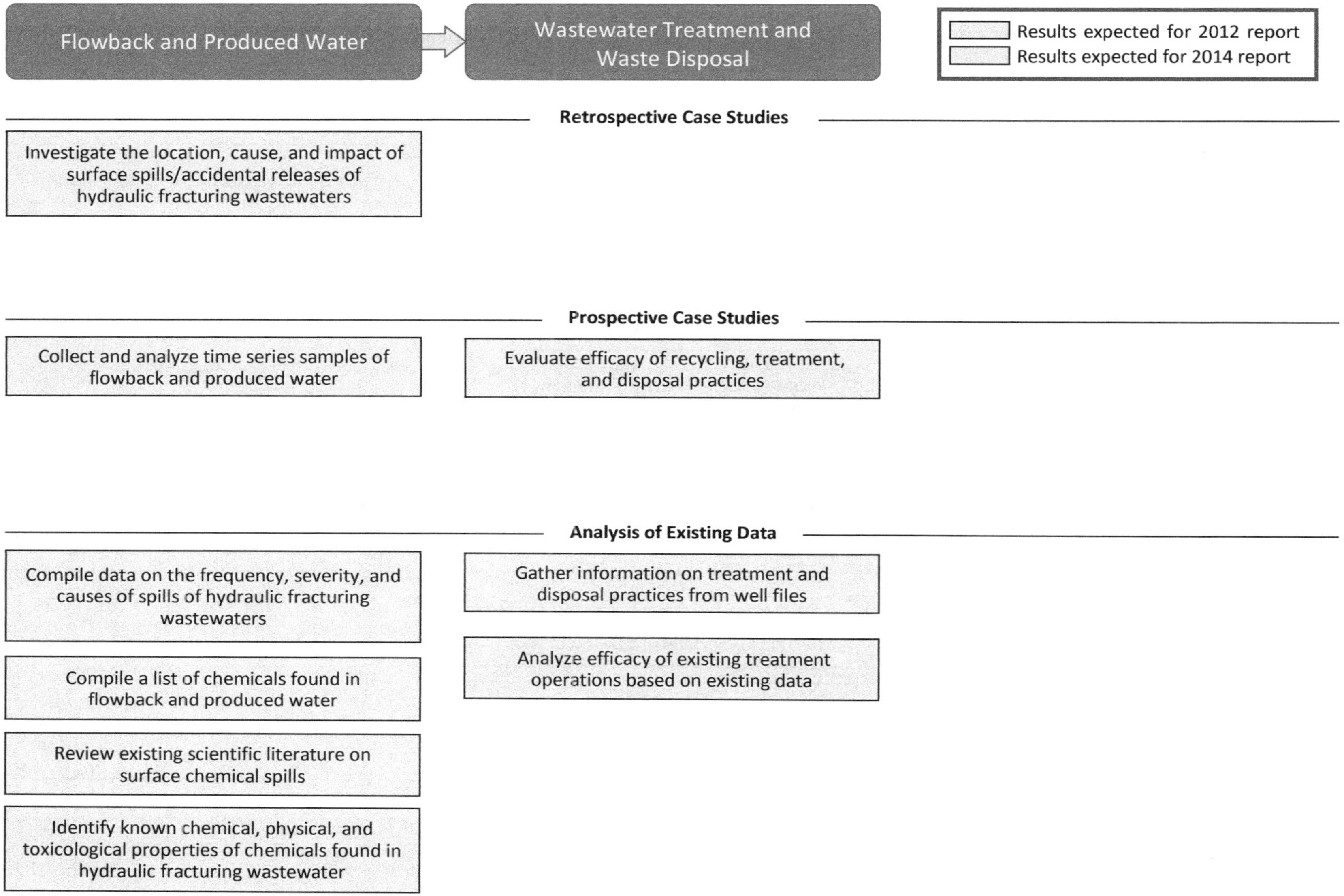

Flowback and Produced Water

Wastewater Treatment and Waste Disposal

Results expected for 2012 report
Results expected for 2014 report

Retrospective Case Studies

Investigate the location, cause, and impact of surface spills/accidental releases of hydraulic fracturing wastewaters

Prospective Case Studies

Evaluate efficacy of recycling, treatment, and disposal practices

Collect and analyze time series samples of flowback and produced water

Analysis of Existing Data

Gather information on treatment and disposal practices from well files

Analyze efficacy of existing treatment operations based on existing data

Compile data on the frequency, severity, and causes of spills of hydraulic fracturing wastewaters

Compile a list of chemicals found in flowback and produced water

Review existing scientific literature on surface chemical spills

Identify known chemical, physical, and toxicological properties of chemicals found in hydraulic fracturing wastewater

FIGURE 11A. SUMMARY OF RESEARCH PROJECTS PROPOSED FOR THE LAST TWO STAGES OF THE HYDRAULIC FRACTURING WATER LIFECYCLE

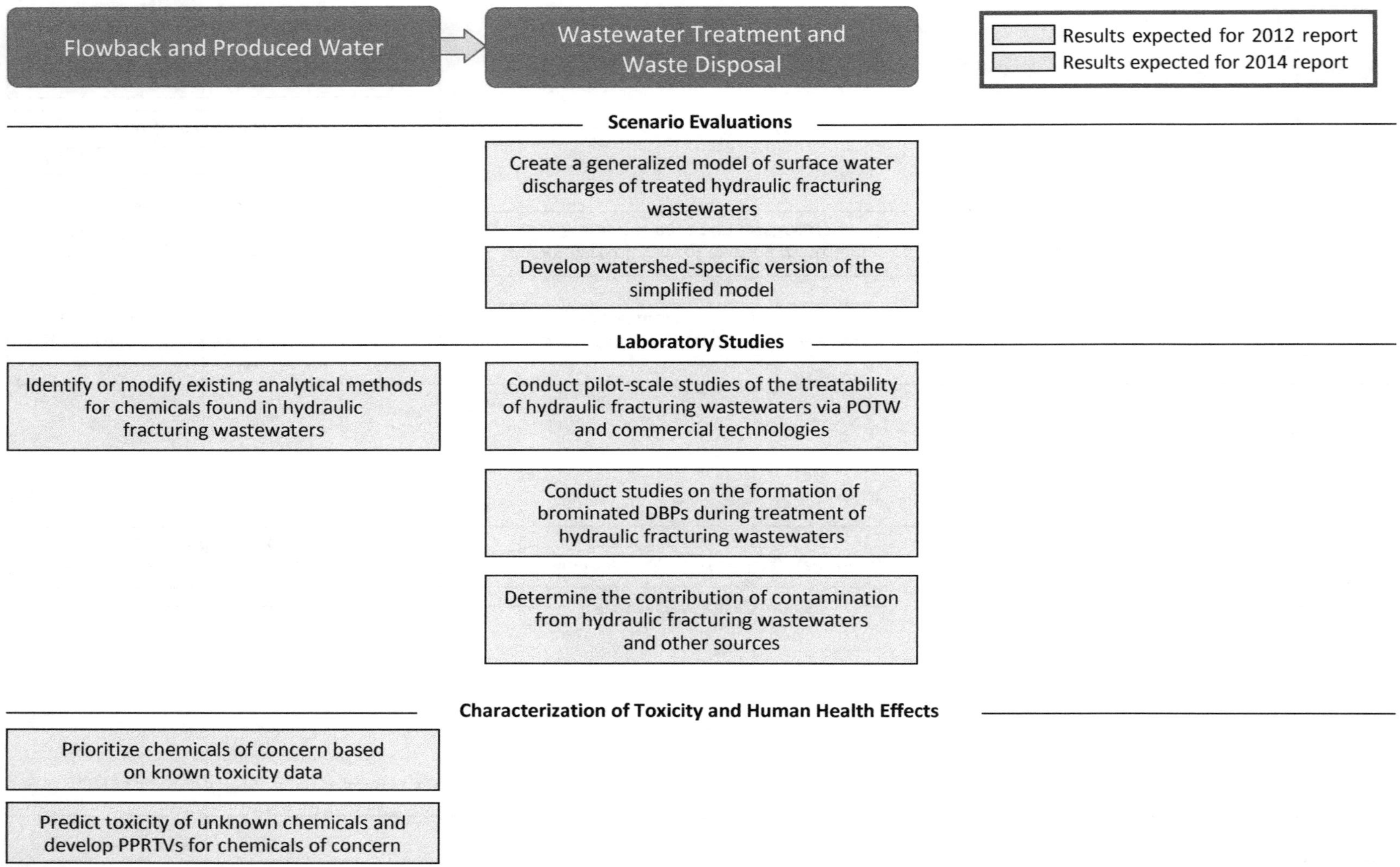

Results expected for 2012 report
Results expected for 2014 report

Flowback and Produced Water

Wastewater Treatment and Waste Disposal

Scenario Evaluations

Create a generalized model of surface water discharges of treated hydraulic fracturing wastewaters

Develop watershed-specific version of the simplified model

Laboratory Studies

Conduct pilot-scale studies of the treatability of hydraulic fracturing wastewaters via POTW and commercial technologies

Conduct studies on the formation of brominated DBPs during treatment of hydraulic fracturing wastewaters

Determine the contribution of contamination from hydraulic fracturing wastewaters and other sources

Identify or modify existing analytical methods for chemicals found in hydraulic fracturing wastewaters

Characterization of Toxicity and Human Health Effects

Prioritize chemicals of concern based on known toxicity data

Predict toxicity of unknown chemicals and develop PPRTVs for chemicals of concern

FIGURE 11B. SUMMARY OF RESEARCH PROJECTS PROPOSED FOR THE LAST TWO STAGES OF THE HYDRAULIC FRACTURING WATER LIFECYCLE

Brief summaries of how the research activities described in Chapter 6 will answer the fundamental research questions appear below:

Water Acquisition: What are the potential impacts of large volume water withdrawals from ground and surface waters on drinking water resources?

The 2012 report will provide a partial answer to this question based on the analysis of existing data. This will include data collected from two information requests and from existing data collection efforts in the Susquehanna River Basin and Garfield County, Colorado. The requested data from hydraulic fracturing service companies and oil and gas operators will provide EPA with general information on the source, quality, and quantity of water used for hydraulic fracturing operations. Data gathered in the Susquehanna River Basin and Garfield County, Colorado, will allow EPA to assess the impacts of large volume water withdrawals in a semi-arid and humid region by comparing water quality and quantity data in areas with no hydraulic fracturing activity to areas with intense hydraulic fracturing activities.

Additional work will be reported in the 2014 report. EPA expects to provide information on local water quality and quantity impacts, if any, that are associated with large volume water withdrawals at the two prospective case study locations: Washington County, Pennsylvania, and DeSoto Parish, Louisiana. These two locations will provide information on impacts from surface (Washington County) and ground (DeSoto Parish) water withdrawals for hydraulic fracturing. The site-specific data can then be compared to future scenario modeling of cumulative hydraulic fracturing-related water withdrawals in the Susquehanna River Basin and Garfield County, Colorado, which will model the long-term impacts of multiple hydraulically fractured oil and gas wells within a single watershed. EPA will use the futures scenarios to assess the sustainability of hydraulic fracturing activities in semi-arid and humid environments and to determine what factors (e.g., droughts) may affect predicted impacts.

Chemical Mixing: What are the possible impacts of surface spills on or near well pads of hydraulic fracturing fluids on drinking water resources?

In general, EPA expects to be able to provide information on the composition hydraulic fracturing fluids and summarize the frequency, severity, and causes of spills of hydraulic fracturing fluids in the 2012 report. EPA will use the information gathered from nine hydraulic fracturing service operators to summarize the types of hydraulic fracturing fluids, their composition, and a description of the factors that may determine which chemicals are used. The 2012 report will also provide a list of chemicals used in hydraulic fracturing fluids and their known or predicted chemical, physical, and toxicological properties. Based on known or predicted properties, a small fraction of these chemicals will be identified as chemicals of concern and will be highlighted for additional toxicological analyses or analytical method development, if needed. EPA will use this chemical list to identify available research on the fate and transport of hydraulic fracturing fluid chemical additives in environmental media.

The 2014 report will contain results of additional toxicological analyses of hydraulic fracturing fluid chemical additives with little or no known toxicological data. PPRTVs may be developed for high priority chemicals of concern. EPA will also include the results of the retrospective case study investigations. These investigations will provide verification of whether contamination of drinking water resources has occurred, and if so, if a surface spill of hydraulic fracturing fluids could be responsible for the contamination.

Well Injection: What are the possible impacts of the injection and fracturing process on drinking water resources?

In 2012, EPA will primarily report on the results of the well file analysis and scenario evaluations to assess the role that the mechanical integrity of the wells and existing geologic/man-made features may play in the contamination of drinking water resources due to hydraulic fracturing. The well file analysis will provide nationwide background information on the frequency and severity of well failures in hydraulically fractured oil and gas wells, and will identify any contributing factors that may have led to these failures. Additionally, the well file analysis will provide information on the types of local geologic or man-made features that industry seeks to characterize prior to hydraulic fracturing, and whether or not these features were found to interact with hydraulic fractures. In a separate effort, EPA will use computer modeling to explore various contamination pathway scenarios involving improper well construction, mechanical integrity failure, and the presence of local geologic/man-made features.

Results presented in the 2014 report will focus primarily on retrospective and prospective case studies and laboratory studies. The case studies will provide information on the methods and tools used to protect and isolate drinking water from oil and gas resources before and during hydraulic fracturing. In particular, the retrospective case studies may offer information on the impacts to drinking water resources from failures in well construction or mechanical integrity. EPA will use samples of the shale formations obtained at prospective case study locations to investigate geochemical reactions between hydraulic fracturing fluids and the natural gas-containing formation. These studies will be used to identify important biogeochemical reactions between hydraulic fracturing fluids and environmental media and whether this interaction may lead to the mobilization of naturally occurring materials. By evaluating chemical, physical, and toxicological characteristics of those substances, EPA will be able to determine which naturally occurring materials may be of most concern for human health.

Flowback and Produced Water: What are the possible impacts of surface spills on or near well pads of flowback and produced water on drinking water resources?

EPA will use existing data to summarize the composition of flowback and produced water, as well as what is known about the frequency, severity, and causes of spills of hydraulic fracturing wastewater. Based on information submitted by the hydraulic fracturing service companies and oil and gas operators, EPA will compile a list of chemical constituents found in hydraulic fracturing wastewaters and the factors that may influence this composition. EPA will then use existing databases to determine the chemical, physical, and toxicological properties of wastewater constituents, and will identify specific constituents that may be of particular concern due to their mobility, toxicity, or production volumes. Properties of chemicals with little or no existing information will be estimated using QSAR methods, and high-priority chemicals with no existing toxicological information may be flagged for further analyses. The list of hydraulic fracturing wastewater constituents will also be used as a basis for a review of existing scientific literature to determine the fate and transport of these chemicals in the environment. These results, in combination with the above data analysis, will be presented in the 2012 report.

Results from the retrospective and prospective case studies will be presented in the 2014 report. The retrospective case studies will involve investigations of reported drinking water contamination at locations near reported spills of hydraulic fracturing wastewaters. EPA will first verify if contamination of the drinking water resources has occurred, and if so, then identify the source of this contamination. This may or may not be due to spills of hydraulic fracturing wastewaters. These case studies may provide EPA with information on the impacts of spills of hydraulic fracturing wastewaters to nearby drinking water resources. Prospective case studies will give EPA the opportunity to collect and analyze samples of flowback and produced water at different times, leading to a better understanding of the variability in the composition of these wastewaters.

Wastewater Treatment and Waste Disposal: What are the possible impacts of inadequate treatment of hydraulic fracturing wastewaters on drinking water resources?

In the 2012 report, EPA will analyze existing data, the results from scenario evaluations and laboratory studies to assess the treatment and disposal of hydraulic fracturing wastewaters. Data provided by oil and gas operators will be used to better understand common treatment and disposal methods and where these methods are practiced. This understanding will inform EPA's evaluation of the efficacy of current treatment processes. In a separate effort, EPA researchers will create a generalized computer model of surface water discharges of treated hydraulic fracturing wastewaters. The model will be used to determine the potential impacts of these wastewaters on the operation of drinking water treatment facilities.

Research presented in the 2014 report will include the results of laboratory studies of current treatment and disposal technologies, building upon the results reported in 2012. These studies will provide information on fate and transport processes of hydraulic fracturing wastewater contaminants during treatment by a wastewater treatment facility. Additional laboratory studies will be used to determine the extent of brominated DBP formation in hydraulic fracturing wastewaters, either from brominated chemical additives or high bromide concentrations. If possible, EPA will also collect samples of wastewater treatment plant discharges and stream/river samples to determine the contribution of treated hydraulic fracturing wastewater discharges to stream/river contamination. The generalized computer model described above will be expanded to develop a watershed-specific version that will provide additional information on potential impacts to drinking water intakes and what factors may influence these impacts.

The results for each individual research project will be made available to the public after undergoing a comprehensive quality assurance review. Figures 10 and 11 show which parts of the research will be completed in time for the 2012 report and which components of the study plan are expected to be completed for the 2014 report. Both reports will use the results of the research projects to assess the impacts, if any, of hydraulic fracturing on drinking water resources. Overall, this study will provide data on the key factors in the potential contamination of drinking water resources as well as information about the toxicity of chemicals associated with hydraulic fracturing. The results may then be used in the future to inform a more comprehensive assessment of the potential risks associated with exposure to contaminants associated with hydraulic fracturing activities in drinking water.

Conclusion

This study plan represents an important milestone in responding to the direction from the US Congress in Fiscal Year 2010 to conduct research to examine the relationship between hydraulic fracturing and drinking water resources. EPA is committed to conducting a study that uses the best available science, independent sources of information, and a transparent, peer-reviewed process that will ensure the validity and accuracy of the results. The Agency will work in consultation with other federal agencies, state and interstate regulatory agencies, industry, non-governmental organizations, and others in the private and public sector in carrying out the study. Stakeholder outreach as the study is being conducted will continue to be a hallmark of our efforts, just as it was during the development of this study plan.

13 ADDITIONAL RESEARCH NEEDS

Although EPA's current study focuses on potential impacts of hydraulic fracturing on drinking water resources, stakeholders have identified additional research areas related to hydraulic fracturing operations, as discussed below. Integrating the results of future work in these areas with the findings of the current study would provide a comprehensive view of the potential impacts of hydraulic fracturing on human health and the environment. If opportunities arise to address these concerns, EPA will include them in this current study as they apply to potential impacts of hydraulic fracturing on drinking water resources. However, the research described in this study plan will take precedence.

13.1 USE OF DRILLING MUDS IN OIL AND GAS DRILLING

Drilling muds are known to contain a wide variety of chemicals that might impact drinking water resources. This concern is not unique to hydraulic fracturing and may be important for oil and gas drilling in general. The study plan is restricted to specifically examining the hydraulic fracturing process and will not evaluate drilling muds.

13.2 LAND APPLICATION OF FLOWBACK OR PRODUCED WATERS

Land application of wastewater is a fairly common practice within the oil and gas industry. EPA plans to identify hydraulic fracturing-related chemicals that may be present in treatment residuals. However, due to time constraints, land application of hydraulic fracturing wastes and disposal practices associated with treatment residuals is outside the scope of the current study.

13.3 IMPACTS FROM DISPOSAL OF SOLIDS FROM WASTEWATER TREATMENT PLANTS

In the process of treating wastewater, the solids are separated from the liquid in the mixture. The handling and disposal of these solids can vary greatly before they are deposited in pits or undergo other disposal techniques. These differences can greatly affect exposure scenarios and the toxicological characteristics of the solids. For this reason, a comprehensive assessment of solids disposal is beyond the current study's resources. However, EPA will use laboratory-scale studies to focus on determining the fate and transport of hydraulic fracturing water contaminants through wastewater treatment processes, including partitioning in treatment residuals.

13.4 DISPOSAL OF HYDRAULIC FRACTURING WASTEWATERS IN CLASS II UNDERGROUND INJECTION WELLS

Particularly in the West, millions of gallons of produced water and flowback are transported to Class II UIC wells for disposal. This study plan does not propose to evaluate the potential impacts of this regulated practice or the associated potential impacts due to the transport and storage leading up to ultimate disposal in a UIC well.

13.5 FRACTURING OR RE-FRACTURING EXISTING WELLS

In addition to concerns related to improper well construction and well abandonment processes, there are concerns about the repeated fracturing of a well over its lifetime. Hydraulic fracturing can be repeated as necessary to maintain the flow of hydrocarbons to the well. The near- and long-term effects of repeated pressure treatments on well construction components (e.g., casing and cement) are not well understood. While EPA recognizes that fracturing or re-fracturing existing wells should also be considered for potential impacts to drinking water resources, EPA has not been able to identify potential partners for a case study; therefore, this practice is not considered in the current study. The issues of well age, operation, and maintenance are important and warrant more study.

13.6 COMPREHENSIVE REVIEW OF COMPROMISED WASTE CONTAINMENT

Flowback is deposited in pits or tanks available on site. If these pits or tanks are compromised by leaks, overflows, or flooding, flowback can potentially affect surface and ground water. This current study partially addresses this issue. EPA will evaluate information on spills collected from incident reports submitted by hydraulic fracturing service operators and observations from the case studies. However, a thorough review of pit or storage tank containment failures is beyond the scope of this study.

13.7 AIR QUALITY

There are several potential sources of air emissions from hydraulic fracturing operations, including the off-gassing of methane from flowback before the well is put into production, emissions from truck traffic and diesel engines used in drilling equipment, and dust from the use of dirt roads. There have been reports of changes in air quality from natural gas drilling that have raised public concerns. Stakeholders have also expressed concerned over the potential greenhouse gas impacts of hydraulic fracturing. This study plan does not propose to address the potential impacts from hydraulic fracturing on air quality or greenhouse gases because these issues fall outside the scope of assessing potential impacts on drinking water resources.

13.8 TERRESTRIAL AND AQUATIC ECOSYSTEM IMPACTS

Stakeholders have expressed concern that hydraulic fracturing may have effects on terrestrial and aquatic ecosystems unrelated to its effects on drinking water resources. For example, there is concern that contamination from chemicals used in hydraulic fracturing could result either from accidents during their use, transport, storage, or disposal; spills of untreated wastewater; or planned releases from wastewater treatment plants. Other impacts could result from increases in vehicle traffic associated with hydraulic fracturing activities, disturbances due to site preparation and roads, or stormwater runoff from the drilling site. This study plan does address terrestrial and aquatic ecosystem impacts from hydraulic fracturing because this issue is largely outside the scope of assessing potential impacts on drinking water resources.

13.9 SEISMIC RISKS

It has been suggested that drilling and/or hydraulically fracturing shale gas wells might cause low-magnitude earthquakes. Public concern about this possibility has emerged due to several incidences where weak earthquakes have occurred in several locations with recent increases in drilling, although no conclusive link between hydraulic fracturing and these earthquakes has been found. The study plan does not propose to address seismic risks from hydraulic fracturing, because they are outside the scope of assessing potential impacts on drinking water resources.

13.10 OCCUPATIONAL RISKS

Occupational risks are of concern in the oil and gas extraction industry in general. For example, NIOSH reports that the industry has an annual occupational fatality rate eight times higher than the rate for all US workers, and that fatality rates increase when the level of drilling activity increases (NIOSH, 2009). Acute and chronic health effects associated with worker exposure to hydraulic fracturing fluid chemicals could be of concern. Exposure scenarios could include activities during transport of materials, chemical mixing, delivery, and any potential accidents. The nature of this work poses potential risks to workers that have not been well characterized. Therefore, the recent increase in gas drilling and hydraulic fracturing activities may be a cause for concern with regard to occupational safety. The study plan does not propose to address occupational risks from hydraulic fracturing, because this issue is outside the scope of assessing potential impacts on drinking water resources.

13.11 PUBLIC SAFETY CONCERNS

Emergency situations such as blowouts, chemical spills from sites with hydraulic fracturing, or spills from the transportation of materials associated with hydraulic fracturing (either to or from the well pad) could potentially jeopardize public safety. Stakeholders also have raised concerns about the possibility of public safety hazards as a result of sabotage and about the need for adequate security at drilling sites. This issue is not addressed in the study plan because it is outside the scope of assessing potential impacts on drinking water resources.

13.12 Economic Impacts

Some stakeholders value the funds they receive for allowing drilling and hydraulic fracturing operations on their properties, while others look forward to increased job availability and more prosperous businesses. It is unclear, however, what the local economic impacts of increased drilling activities are and how long these impacts may last. For example, questions have been raised concerning whether the high-paying jobs associated with oil and gas extraction are available to local people, or if they are more commonly filled by those from traditional oil and gas states who have specific skills for the drilling and fracturing process. It is important to better understand the benefits and costs of hydraulic fracturing operations. However, the study plan does not address this issue, because it is outside the scope of assessing potential impacts on drinking water resources

13.13 Sand Mining

As hydraulic fracturing operations have become more prevalent, the demand for proppants has also risen. This has created concern over increased sand mining and associated environmental effects. Some stakeholders are worried that sand mining may lower air quality, adversely affect drinking water resources, and disrupt ecosystems (Driver, 2011). The impact of sand mining should be studied in the future, but is outside the scope of the current study because it falls outside the hydraulic fracturing water lifecycle framework established for this study.

REFERENCES

API (American Petroleum Institute). (2009a, July). *Environmental protection for onshore oil and gas production operations and leases.* API Recommended Practice 51R, first edition. Washington, DC: American Petroleum Institute. Retrieved June 24, 2011, from http://www.api.org/plicy/exploration/hydraulicfracturing/upload/API_RP_S1R.pdf

API (American Petroleum Institute). (2009b, October). *Hydraulic fracturing operations—well construction and integrity guidelines.* API Guidance Document HF1. Washington, DC: American Petroleum Institute.

API (American Petroleum Institute). (2010a, June). *Water management associated with hydraulic fracturing.* API Guidance Document HF2, first edition. Washington, DC: American Petroleum Institute. Retrieved January 20, 2011, from http://www.api.org/Standards/new/api-hf2.cfm.

API (American Petroleum Institute). (2010b, July 19). *Freeing up energy—hydraulic fracturing: Unlocking America's natural gas resources.* Washington, DC: American Petroleum Institute. Retrieved December 2, 2010, from http://www.api.org/policy/exploration/hydraulicfracturing/upload/ HYDRAULIC_FRACTURING_PRIMER.pdf.

Armstrong, K., Card, R., Navarette, R., Nelson, E., Nimerick, K., Samuelson, M., Collins, J., Dumont, G., Priaro, M., Wasylycia, N., & Slusher, D. (1995, Autumn). Advanced fracturing fluids improve well economics. *Oil Field Review,* 34-51.

Arthur, J. D., Bohm, B., & Layne, M. (2008, September 21-24). *Hydraulic fracturing considerations for natural gas wells of the Marcellus Shale.* Presented at The Ground Water Protection Council 2008 Annual Forum, Cincinnati, OH.

Baker Hughes. (2010, June 11). *Baker Hughes rig count blog.* Retrieved August 10, 2010, from http://blogs.bakerhughes.com/rigcount.

Bellabarba, M., Bulte-Loyer, H., Froelich, B., Le Roy-Delage, S., Kujik, R., Zerouy, S., Guillot, D., Meroni, N., Pastor, S., & Zanchi, A. (2008, Spring). Ensuring zonal isolation beyond the life of the well. *Oil Field Review,* 18-31.

Berman, A. (2009, August 1). Lessons from the Barnett Shale suggest caution in other shale plays. *World Oil, 230*(8).

Blauch, M. (2011, March 29). *Shale frac sequential flowback analyses and reuse implications.* Presented at the EPA's Hydraulic Fracturing Technical Workshop 4, Washington, DC.

Breit, G.N. (2002). Produced waters database: US Geological Survey. Accessed September 20, 2011 from http://energy.cr.usgs.gov/prov/prodwat/index.htm.

Bryant, J., Welton, T., & Haggstrom, J. (2010, September 1). Will flowback or produced water do? *E&P.* Retrieved January 19, 2011, from http://www.epmag.com/Magazine/2010/9/item65818.php.

Carter, R. H., Holditch, S. A., & Wolhart, S. L. (1996, October 6-9). *Results of a 1995 hydraulic fracturing survey and a comparison of 1995 and 1990 industry practices.* Presented at the Society of Petroleum Engineers Annual Technical Conference, Denver, CO.

Castle, J. W., Falta, R. W., Bruce, D., Murdoch, L., Foley, J., Brame, S. E., & Brooks, D. (2005). *Fracture dissolution of carbonate rock: an innovative process for gas storage.* Topical Report, DOE, NETL, DE-FC26-02NT41299. Washington, DC: Department of Energy.

Chapra, S.C. (2008). *Surface water quality modeling.* Long Grove, IL: Waveland Press.

Chesapeake Energy. (2009). *Barnett Shale—natural gas production.* Retrieved August 9, 2010, from http://www.askchesapeake.com/Barnett-Shale/Production/Pages/information.aspx.

Chesapeake Energy. (2010, July). *Hydraulic fracturing fact sheet.* Retrieved August 9, 2010, from http://www.chk.com/Media/CorpMediaKits/Hydraulic_Fracturing_Fact_Sheet.pdf.

Cipolla, C. L., & Wright, C. A. (2000, April 3-5). *Diagnostic techniques to understand hydraulic fracturing: What? Why? And how?* Presented at the Society of Petroleum Engineers/Canadian Energy Research Institute Gas Technology Symposium, Calgary, Alberta, Canada.

Clark, C. E, & Veil, J. A. (2009). *Produced water volumes and management practices in the US* Washington, DC: US Department of Energy, National Energy Technology Laboratory, Project No. DE-AC02-06CH11357. Retrieved July 27, 2010, from http://www.netl.doe.gov/technologies/coalpower/ewr/water/pdfs/anl%20produced%20water%20volumes%20sep09.pdf.

Daneshy, A. A. (2003, April). Off-balance growth: A new concept in hydraulic fracturing. No. SPE 80992. *Journal of Petroleum Technology (Distinguished Author Series), 55*(4), 78-85.

Doherty, J. (2011, July-August). Modeling: Picture perfect or abstract art? *Ground Water, 49*(4), 455.

Driver, A. (2011, September 21). Critics of energy 'fracking' raise new concern: sand. Reuters. Retrieved September 22, 2011, from http://www.msnbc.msn.com/id/44612454/ns/us_news-environment/t/critics-energy-fracking-raise-new-concern-sand/.

Eby, G. N. (2004). *Principles of environmental geochemistry.* Pacific Grove, CA: Thompson-Brooks/Cole.

Falk, H., Lavergren, U., & Bergback, B. (2006). Metal mobility in alum shale from Öland, Sweden. *Journal of Geochemical Exploration, 90*(3), 157-165.

Gadd, G. M. (2004). Microbial influences on metal mobility and application for bioremediation. *Geoderma, 122*, 109-119.

Galusky, L. P., Jr. (2007, April 3). *Fort Worth Basin/Barnett Shale natural gas play: An assessment of present and projected fresh water use.* Fort Worth, TX: Barnett Shale Water Conservation and Management Committee. Retrieved July 21, 2010, from www.barnettshalewater.org/uploads/Barnett_Water_Availability_Assessment__Apr_3__2007.pdf.

Gaudlip, A. W., & Paugh, L. O. (2008, November 18). *Marcellus Shale water management challenges in Pennsylvania* (No. SPE 119898). Presented at the Society of Petroleum Engineers Shale Gas Production Conference, Irving, TX.

Godsey, W.E. (2011, March 29). *Fresh, brackish, or saline water for hydraulic fracs: What are the options?* Presented at the EPA's Hydraulic Fracturing Technical Workshop 4, Washington, DC.

GWPC (Ground Water Protection Council). (2009). *State oil and natural gas regulations designed to protect water resources.* Washington, DC: US Department of Energy, National Energy Technology Laboratory. Retrieved July 23, 2010, from http://data.memberclicks.com/site/coga/GWPC.pdf.

GWPC (Ground Water Protection Council) & ALL Consulting. (2009). *Modern shale gas development in the US: A primer.* Contract DE-FG26-04NT15455. Washington, DC: US Department of Energy, Office of Fossil Energy and National Energy Technology Laboratory. Retrieved August 2, 2010, from http://www.netl.doe.gov/technologies/oil-gas/publications/EPreports/ Shale_Gas_Primer_2009.pdf.

Halliburton. (2008). *US shale gas – an unconventional resource, unconventional challenge.* Retrieved September 7, 2011, from http://www.halliburton.com/public/solutions/contents/Shale/related_docs/H063771.pdf.

Hall, B. E., & Larkin, S. D. (1989). On-site quality control of fracture treatments. *Journal of Petroleum Technology, 41*(5), 526-532.

Hanson, G. (2011, March 29). *How are appropriate water sources for hydraulic fracturing determined? Pre-development conditions and management of development phase water usage.* Presented at the EPA's Hydraulic Fracturing Technical Workshop 4, Washington, DC.

Harper, J. A. (2008). The Marcellus Shale—An old "new" gas reservoir in Pennsylvania. *Pennsylvania Geology, 38*(1), 2-13.

Hayes, T. (2009a, June 4). *Gas shale produced water.* Presented at the Research Partnership to Secure Energy for America/Gas Technology Institute Gas Shales Forum, Des Plaines, IL. Retrieved August 11, 2010, from http://www.rpsea.org/attachments/contentmanagers/429/Gas_Shale_Produced_Water_-_Dr._Tom_Hayes_GTI.pdf.

Hayes, T. (2009b, December 31). *Sampling and analysis of water streams associated with the development of Marcellus Shale gas, final report.* Canonsburg, PA: Marcellus Shale Coalition, Gas Technology Institute.

Hayes, T. (2011, March 29). *Characterization of Marcellus shale and Barnett shale flowback waters and technology development for water reuse.* Presented at the EPA's Hydraulic Fracturing Technical Workshop 4, Washington, DC.

Holditch, S. A. (1993, March). Completion methods in coal-seam reservoirs. *Journal of Petroleum Technology, 45*(3), 270-276.

Hopey, D. (2011, March 5). Radiation-fracking link sparks swift reactions. *Pittsburgh Post-Gazette*. Retrieved August 31, 2011, from http://www.post-gazette.com/pg/11064/1129908-113.stm.

Hopey, D., & Hamill, S.D. (2011, April 19). Pa.: Marcelus wastewater shouldn't go to treatment plants. *Pittsburgh Post-Gazette*. Retrieved August 31, 2011, from http://www.post-gazette.com/pg/11109/1140412-100-0.stm.

Horn, A. D. (2009, March 24). *Breakthrough mobile water treatment converts 75% of fracturing flowback fluid to fresh water and lowers CO_2 emissions* (No. SPE 121104). Presented at the Society of Petroleum Engineers E&P Environmental and Safety Conference, San Antonio, TX.

Hossain, Md. M., & Rahman, M. K. (2008). Numerical simulation of complex fracture growth during tight reservoir stimulation by hydraulic fracturing. *Journal of Petroleum Science and Engineering, 60*, 86-104.

ICF International. (2009a, August 5). *Technical assistance for the draft supplemental generic EIS: oil, gas and solution mining regulatory program. Well permit issuance for horizontal drilling and high-volume hydraulic fracturing to develop the Marcellus Shale and other low permeability gas reservoirs—Task 2.* Albany, NY: ICF Incorporated, LLC, New York State Energy Research and Development Authority Contract PO Number 9679. Retrieved July 25, 2010, from http://www.nyserda.org/publications/ ICF%20Task%202%20Report_Final.pdf.

ICF International. (2009b, August 7). *Technical assistance for the draft supplemental generic EIS: oil, gas and solution mining regulatory program. Well permit issuance for horizontal drilling and high-volume hydraulic fracturing to develop the Marcellus Shale and other low permeability gas reservoirs—Task 1.* Albany, NY: ICF Incorporated, LLC , New York State Energy Research and Development Authority Contract PO Number 9679. Retrieved July 25, 2010, from http://www.nyserda.com/ publications/ICF%20Task%201%20Report_Final.pdf.

Jeu, S. J., Logan, T. L., & McBane, R. A. (1988, October 2-5). *Exploitation of deeply buried coalbed methane using different hydraulic fracturing techniques in the Piceance Basin, Colorado, and San Juan Basin, New Mexico.* Presented at the Society of Petroleum Engineers Annual Technical Conference and Exhibition, Houston, TX.

Jobson, H.E. (1996). *Prediction of traveltime and longitudinal dispersion in rivers and streams*. ISGS Water-Resources Investigations, Report 96-4013.

Judson, R. S., Martin, M. T., Reif, D. M., Houck, K. A., Knudsen, T. B., Rotroff, D. M., Xia, M., Sakamuru, S., Huang, R., Shinn, P., Austin, C. P., Kavlock, R. J., & Dix, D. J. (2010a). Analysis of eight oil spill dispersants using rapid, *in vitro* tests for endocrine and other biological activity. *Environmental Science & Technology, 44*, 5979-5985.

Judson, R. S., Houck, K. A., Kavlock, R. J., Knudsen, T. B., Martin, M. T., Mortensen, H. M., Reif, D. M., Rotroff, D. M., Shah, I., Richard, A. M., & Dix, D. J. (2010b). *In vitro* screening of environmental chemicals for targeted testing prioritization: The ToxCast project. *Environmental Health Perspectives*, *118*, 485-492.

Kargbo, D. M., Wilhelm, R. G., & Campbell, D. J. (2010). Natural gas plays in the Marcellus Shale: challenges and potential opportunities. *Environmental Science & Technology, 44*(15), 5679-5684.

Keister, T. (2009, January 12). *Marcellus gas well water supply and wastewater disposal, treatment, and recycle technology*. Brockway, PA: ProChemTech International, Inc. Retrieved July 29, 2010, from http://www.prochemtech.com/Literature/TAB/PDF_TAB_Marcellus_Gas_Well_Water_Recycle.pdf.

Kellman, S., & Schneider, K. (2010, September 15). Water demand is flash point in Dakota oil boom. *Circle of Blue Waternews*. Retrieved September 18, 2010, from http://www.circleofblue.org/waternews/2010/world/scarce-water-is-no-limit-yet-to-north-dakota-oil-shale-boom/.

Konikow, L.F. (2010). The secret to successful solute-transport modeling. *Groundwater, 49*(2), 144-159.

Lee, J.J. (2011a, March 29). *Water quality in the development area of the Marcellus shale gas in Pennsylvania and the implications on discerning impacts from hydraulic fracturing.* Presented at the EPA's Hydraulic Fracturing Technical Workshop 4, Washington, DC.

Lee, J.J. (2011b, March 30). *Hydraulic fracturing and safe drinking water*. Presented at the EPA's Hydraulic Fracturing Technical Workshop 4, Washington, DC.

Lee, M. (2011, April 20). Chesapeake battles out-of-control Marcellus gas well. Bloomberg. Retrieved August 31, 2011, from http://www.bloomberg.com/news/2011-04-20/chesapeake-battles-out-of-control-gas-well-spill-in-pennsylvania.html.

Legere, L. (2011, August 13). State pushes for legal end to shale wastewater discharges. *The Times Tribune*. Retrieved August 31, 2011, from http://thetimes-tribune.com/news/state-pushes-for-legal-end-to-shale-wastewater-discharges-1.1188211#axzz1VDXItBd1.

Leventhal, J. S., & Hosterman, J. W. (1982). Chemical and mineralogical analysis of Devonian black shale samples from Martin County, Kentucky; Caroll and Washington Counties, Ohio; Wise County, Virginia; and Overton County, Tennessee. *Chemical Geology, 37,* 239-264.

Long, D. T., & Angino, E. E. (1982). The mobilization of selected trace metals from shales by aqueous solutions: Effects of temperature and ionic strength. *Economic Geology, 77*(3), 646-652.

Louisiana Office of Conservation. (2011, August 19). *Order No. ENV 2011-GW014*. Retrieved October 19, 2011, from http://dnr.louisiana.gov/assets/news_releases/OrderENV2011-GW0140001.pdf.

Lustgarten, A. (2009, September 21). Frack fluid spill in Dimock contaminates stream, killing fish. ProPublica. Retrieved August 31, 2011, from http://www.propublica.org/article/frack-fluid-spill-in-dimock-contaminates-stream-killing-fish-921.

Maclin, E., Urban, R., & Haak, A. (2009, December 31). *Re: New York State Department of Environmental Conservation's draft supplemental generic environmental impact statement on the oil, gas, and solution mining regulatory program.* Arlington, VA: Trout Unlimited. Retrieved July 26, 2010, from http://www.tcgasmap.org/media/ Trout%20Unlimited%20NY%20Comments%20on%20Draft%20SGEIS.pdf.

Martin, T., & Valkó, P. (2007). Hydraulic fracture design for production enahancement. In M.J. Economides & T. Martin (Eds.), *Modern Fracturing: Enhancing Natural Gas Production* (p95) ET Publishing, Houston, TX.

McLean, J. S., & Beveridge, T. J. (2002). Interactions of bacteria and environmental metals, fine-grained mineral development, and bioremediation strategies. In P. M. Haung, et al. (Eds.), *Interactions between soil particles and microrganisms* (pp. 67-86). New York, NY: Wiley.

McMahon, P. B., Thomas, J. C., & Hunt, A. G. (2011). *Use of diverse geochemical data sets to determine sources and sinks of nitrate and methane in groundwater, Garfield County, Colorado, 2009.* US Geological Survey Scientific Investigations Report 2010–5215. Reston, VA: US Department of the Interior, US Geological Survey.

Myers, T. (2009). *Technical memorandum: Review and analysis of draft supplemental generic environmental impact statement on the oil, gas and solution mining regulatory program. Well permit issuance for horizontal drilling and high-volume hydraulic fracturing to develop the Marcellus Shale and other low-permeability gas reservoirs.* New York, NY: Natural Resources Defense Council. Retrieved July 26, 2010, from http://www.tcgasmap.org/media/NRDCMyers%20Comments%20on%20Draft% 20SGEIS.pdf.

National Research Council. (2010). *Management and effects of coalbed methane produced water in the western US.* Washington, DC: National Academies Press.

Nemat-Nassar, S., Abe, H., & Hirakawa, S. (1983). *Hydraulic fracturing and geothermal energy.* The Hague, The Netherlands: Kluwer Academic Publishers.

New Hampshire Department of Environmental Services. (2010). Environmental fact sheet. Well development by hydro-fracking. Concord, NH: New Hampshire Department of Environmental Services. Retrieved January 11, 2011, from http://des.nh.gov/organization/commissioner/pip/factsheets/dwgb/documents/dwgb-1-3.pdf.

NIOSH (National Institute for Occupational Safety and Health). (2009, February). *Oil and gas extraction. Inputs: Occupational safety and health risks.* Atlanta, GA: Centers for Disease Control and Prevention. Retrieved September 17, 2010, from http://www.cdc.gov/niosh/programs/oilgas/risks.html.

NYSDEC (New York State Department of Environmental Conservation). (2011, September). *Supplemental generic environmental impact statement on the oil, gas and solution mining regulatory program (revised draft). Well permit issuance for horizontal drilling and high-volume hydraulic fracturing to develop the Marcellus Shale and other low-permeability gas reservoirs.* Albany, NY: New York State Department of

Environmental Conservation. Retrieved January 20, 2010, from ftp://ftp.dec.state.ny.us/dmn/download/OGdSGEISFull.pdf.

Oil and Gas Investor. (2005, March). *Tight Gas* (special supplement). Houston, TX: Oil and Gas Investor/Hart Energy Publishing LP. Retrieved August 9, 2010, from http://www.oilandgasinvestor.com/pdf/Tight%20Gas.pdf.

OilGasGlossary.com. (2010). *Drilling fluid definition.* Retrieved February 3, 2011, from http://oilgasglossary.com/drilling-fluid.html.

OilShaleGas.com. (2010). OilShaleGas.com—oil & shale gas discovery news. Retrieved January 17, 2011, from http://oilshalegas.com.

Oreskes, N. K., Shrader-Frechette, K., & Belitz, K. (1994, February 4). Verification, validation, and confirmation of numerical models in the earth sciences. *Science, 263*(5147), 641-646.

Oreskes, N. K. (2003). The role of quantitative models in science. In C. D. Canham, J. J. Cole, & W. K. Lauenroth (Eds.), *Models in ecosystem science* (pp. 13-31). Princeton, NJ: Princeton University Press.

Osborn, S.G., Vengosh, A., Warner, N.R., Jackson, R.B. (2011). Methane contamination of drinking water accompanying gas-well drilling and hydraulic fracturing. *Proceedings of the National Academy of Sciences*, 108(20), 8172-8176.

PADEP (Pennsylvania Department of Environmental Protection). (2010a). *Marcellus Shale*. Harrisburg, PA: Pennsylvania Department of Environmental Protection. Retrieved August 9, 2010, from http://www.elibrary.dep.state.pa.us/dsweb/Get/Document-77964/0100-FS-DEP4217.pdf.

PADEP (Pennsylvania Department of Environmental Protection). (2010b, December 15). *Consent order and settlement agreement (Commonwealth of Pennsylvania Department of Environmental Protection and Cabot Oil & Gas Corporation).* PA: Pennsylvania Department of Environmental Protection.

Palisch, T. T., Vincent, M. C., & Handren, P. J. (2008, September 21-24). *Slickwater fracturing—food for thought.* No. 115766-MS. Paper presented at the Society of Petroleum Engineers Annual Technical Conference, Denver, CO.

Palmer, I. D., Fryan, R. T., Tumino, K. A., & Puri, R. (1991, August 12). Water fracs outperform gel fracs in coalbed pilot. *Oil and Gas Journal*, 71-76.

Palmer, I. D., Lambert, S. W., & Spitler, J. L. (1993). Coalbed methane well completions and stimulations. *AAPG Studies in Geology, 38*, 303-341.

Pashin, J. C. (2007). Hydrodynamics of coalbed methane reservoirs in the Black Warrior Basin: Key to understanding reservoir performance and environmental Issues. *Applied Geochemistry, 22*, 2257-2272.

Pearson, C. M. (1989). *US Patent No. 4,845,981,1989. System for monitoring fluids during well stimulation processes*. Washington, DC: US Patent and Trademark Office.

Pennsylvania Environmental Quality Board. (2009, November 7). Proposed Rulemaking [25 PA. CODE CH. 95] wastewater treatment requirements [39 Pa.B. 6467] [Saturday, November 7, 2009]. *The Pennsylvania Bulletin, 39*(45), Doc. No. 09-2065. Retrieved January 21, 2011, from http://www.pabulletin.com/secure/data/vol39/39-45/2065.html.

Pennsylvania State University. (2010). *Marcellus education fact sheet. Water withdrawals for development of Marcellus Shale gas in Pennsylvania: Introduction to Pennsylvania's water resources.* University Park, PA: College of Agricultural Sciences, Pennsylvania State University. Retrieved November 26, 2010, from http://pubs.cas.psu.edu/freepubs/pdfs/ua460.pdf.

Pickett, A. (2009, March). New solutions emerging to treat and recycle water used in hydraulic fracs. *American Oil & Gas Reporter.* Retrieved July 29, 2010, from http://www.aogr.com/index.php/magazine/cover_story_archives/march_2009_cover_story/.

Piggot, A. R., Elsworth, D. (1996). Displacement of formation fluids by hydraulic fracturing. *Geotechnique, 46*(4), 671-681.

Plewa, M.J., Wagner, E.D. (2009). Quantitative Comparative Mammalian Cell Cytotoxicity and Genotoxicity of Selected Classes of Drinking Water Disinfection By-Products. Water Research Foundation, Denver, CO.

Prouty, J. L. (2001). Tight gas in the spotlight. *Gas Technology Institute GasTIPS, 7*(2), 4-10.

Puko, T. (2010, August 7). Drinking water from Mon deemed safe. *The Pittsburgh Tribune-Review.* Retrieved August 31, 2011, from http://www.pittsburghlive.com/x/pittsburghtrib/news/s_693882.html.

Reif, D. M., Martin, M. T., Tan, S. W., Houck, K. A., Judson, R. S., Richard, A. M., Knudsen, T. B., Dix, D. J., & Kavlock, R. J. (2010). Endocrine profiling and prioritization of environmental chemicals using ToxCast data. *Environmental Health Perspectives, 118*, 1714-1720.

Rogers, R. E., Ramurthy, M., Rodvelt, G., & Mullen, M. (2007). *Coalbed methane: Principles and practices.* Third edition. Starkville, MS: Oktibbeha Publishing Co. Retrieved August 2, 2010, from http://www.halliburton.com/public/pe/contents/Books_and_Catalogs/web/CBM/CBM_Book_Intro.pdf.

Rowan, T. M. (2009, September 23-25). *Spurring the Devonian: Methods of fracturing the lower Huron in southern West Virginia and eastern Kentucky.* Presented at the Society for Petroleum Engineers Eastern Regional Meeting, Charleston, WV.

Rowan, E. L., Engle, M. A., Kirby, C. S., & Kraemer, T. F. (2011, September 7). *Radium content of oil- and gas- field produced waters in the northern Appalachian Basin – Summary and discussion of data.* US Geological Survey Scientific Investigations Report 2011-5135.

Ruszka, J. (2007, August 1). Global challenges drive multilateral drilling. *E&P.* Retrieved August 13, 2010, from http://www.epmag.com/archives/features/583.htm.

Satterfield, J., Kathol, D., Mantell, M., Hiebert, F., Lee, R., & Patterson, K. (2008, September 20-24). *Managing water resource challenges in select natural gas shale plays. GWPC Annual Forum.* Oklahoma City, OK: Chesapeake Energy Corporation. Retrieved July 21, 2010, from http://www.gwpc.org/ meetings/forum/2008/proceedings/Ground%20Water%20&%20Energy/SatterfieldWaterEnergy.pdf.

Southam, G. (2000). Bacterial surface-mediated mineral formation. In D. R. Lovely (Ed.), *Environmental Microbe-Metal Interactions* (pp. 257-276). Washington, DC: American Society of Microbiology.

Sparks, D. L. (1995). *Environmental soil chemistry.* San Diego, CA: Academic Press.

Sposito, G. (1989). *The chemistry of soils.* New York, NY: Oxford University Press.

State of Colorado Oil and Gas Conservation Commission. (2009a, October 5). *Bradenhead test report.* OGCC Operator Number 26420, API Number 123-11848. Denver, CO: State of Colorado Oil and Gas Conservation Commission.

State of Colorado Oil and Gas Conservation Commission. (2009b, December 7). *Sundry notice.* OGCC Operator Number 26420, API Number 05-123-11848. Denver, CO: State of Colorado Oil and Gas Conservation Commission.

State of Colorado Oil and Gas Conservation Commission. (2009c, December 17). *Colorado Oil and Gas Conservation Commission approved Wattenberg Bradenhead testing and staff policy.* Letter sent to all oil and gas operators active in the Denver Basin. Denver, CO: State of Colorado Oil and Gas Conservation Commission.

Stumm, W., & Morgan, J. J. (1996). *Chemical equilibria and rates in natural waters.* Third edition. New York, NY: John Wiley & Sons, Inc.

Tonkin, M., & Dougherty, J. (2009). Efficient nonlinear predictive error variance for highly parameterized models. *Water Resources Research, 45.*

Tuttle, M. L. W., Briet, G. N., & Goldhaber, M. B. (2009). Weathering of the New Albany Shale, Kentucky: II. Redistribution of minor and trace elements. *Applied Geochemistry, 24,* 1565-1578.

URS Corporation. (2009, September 16). *Water-related issues associated with gas production in the Marcellus Shale: Additives use, flowback quality and quantities, regulations, on-site treatment, green technologies, alternate water sources, water well-testing.* Prepared for New York State Energy Research and Development Authority, Contract PO No. 10666. Fort Washington, PA: URS Corporation. Retrieved August 2, 2010, from http://www.nyserda.org/publications/02%20Chapter%202%20-%20URS%202009-9-16.pdf.

US House. (2009). *Department of the Interior, Environment, and related agencies Appropriations Act, 2010.* Washington, DC: Conference of Committee, US House. Retrieved September 23, 2011 from http://frwebgate.access.gpo.gov/cgi-bin/getdoc.cgi?dbname=111_cong_reports&docid=f:hr316.111.pdf.

USEIA (US Energy Information Administration). (2010, December). *Annual energy outlook 2011: Early release overview.* Washington, DC: US Department of Energy. Retrieved January 17, 2011, from http://www.eia.gov/forecasts/aeo/.

USEIA (US Energy Information Administration). (2011a). Glossary. Retrieved September 20, 2011, from http://205.254.135.24/tools/glossary/.

USEIA (US Energy Information Administration). (2011b, October 11). *Oil and natural gas drilling on the rise.* Today in Energy. Retrieved October 15, 2011 from http://www.eia.gov/todayinenergy/detail.cfm?id=3430.

USEPA (US Environmental Protection Agency). (2002, November). *Overview of the EPA quality system for environmental data and technology.* No. EPA/240/R-02/003. Washington, DC: US Environmental Protection Agency, Office of Environmental Information. Retrieved January 20, 2011, from http://www.epa.gov/QUALITY/qs-docs/overview-final.pdf.

USEPA (US Environmental Protection Agency). (2004, June). *Evaluation of impacts to underground sources of drinking water by hydraulic fracturing of coalbed methane reservoirs.* No. EPA/816/R-04/003. Washington, DC: US Environmental Protection Agency, Office of Water. Retrieved January 21, 2011, from http://water.epa.gov/type/groundwater/uic/class2/hydraulicfracturing/wells_coalbedmethanestudy.cfm.

USEPA (US Environmental Protection Agency). (2009). *EPA Records Schedule 501, Applied and Directed Scientific Research.* Retrieved September 7, 2011, from http://www.epa.gov/records/policy/schedule/sched/501.htm.

USEPA (US Environmental Protection Agency). (2010a, March). *Scoping materials for initial design of EPA research study on potential relationships between hydraulic fracturing and drinking water resources.* Washington, DC: US Environmental Protection Agency, Office of Research and Development. Retrieved September 16, 2010, from http://yosemite.epa.gov/sab/sabproduct.nsf/0/3B745430D624ED3B852576D400514B76/$File/Hydraulic+Frac+Scoping+Doc+for+SAB-3-22-10+Final.pdf.

USEPA (US Environmental Protection Agency). (2010b, April 23). *Trip report* (EXCO Resources' gas well drilling site, Norris Ferry Road, southern Caddo Parish (Shreveport), LA). Dallas, TX: US Environmental Protection Agency Region 6.

USEPA (US Environmental Protection Agency). (2010c, June). *Advisory on EPA's research scoping document related to hydraulic fracturing.* Washington, DC: US Environmental Protection Agency, Office of the Administrator, Science Advisory Board. Retrieved September 16, 2010, from http://yosemite.epa.gov/sab/sabproduct.nsf/0/CC09DE2B8B4755718525774D0044F929/$File/EPA-SAB-10-009-unsigned.pdf.

USEPA (US Environmental Protection Agency). (2010d, July). *EPA's action development process: Interim guidance on considering environmental justice during the development of an action.* OPEI Regulatory

Development Series. Washington, DC: US Environmental Protection Agency. Retrieved January 17, 2011, from http://www.epa.gov/environmentaljustice/resources/policy/considering-ej-in-rulemaking-guide-07-2010.pdf.

USEPA (US Environmental Protection Agency). (2011a, February). *Draft plan to study the potential impacts of hydraulic fracturing on drinking water resources.* Washington, DC: US Environmental Protection Agency, Office of Research and Development.

USEPA (US Environmental Protection Agency). (2011b, August). *SAB review of EPA's Draft Hydraulic Fracturing Study Plan.* Washington, DC: US Environmental Protection Agency, Office of the Administrator, Science Advisory Board. Retrieved September 7, 2011, from http://yosemite.epa.gov/sab/sabproduct.nsf/0/2BC3CD632FCC0E99852578E2006DF890/$File/EPA-SAB-11-012-unsigned.pdf.

USGS (US Geological Survey). (1999, September). *Naturally occurring radioactive materials (NORM) in produced water and oil field equipment – an issue for the energy industry.* USGS Fact Sheet FS-142-99. Retrieved September 14, 2011, from http://pubs.usgs.gov/fs/fs-0142-99/fs-0142-99.pdf.

USGS (US Geological Survey). (2002, May 29). *Produced waters database.* Reston, VA: US Geological Survey National Center. Retrieved January 17, 2011, from http://energy.cr.usgs.gov/prov/prodwat/data2.htm.

Veil, J. A., Puder, M. G., Elcock, D., & Redweik, R. J. (2004). *A white paper describing produced water from production of crude oil, natural gas, and coal bed methane.* Prepared for the US Department of Energy, National Energy Technology Laboratory. Argonne, IL: Argonne National Laboratory. Retrieved January 20, 2011, from http://www.evs.anl.gov/pub/doc/ProducedWatersWP0401.pdf.

Veil, J. A. (2007, August). *Trip report for field visit to Fayetteville Shale gas wells.* No. ANL/EVS/R-07/4. Prepared for the US Department of Energy, National Energy Technology Laboratory, project no. DE-FC26-06NT42930. Argonne, IL: Argonne National Laboratory. Retrieved July 27, 2010, from http://www.evs.anl.gov/pub/doc/ANL-EVS_R07-4TripReport.pdf.

Veil, J. A. (2010, July). *Final report: Water management technologies used by Marcellus Shale gas producers.* Prepared for the US Department of Energy, National Energy Technology Laboratory, Department of Energy award no. FWP 49462. Argonne, IL: Argonne National Laboratory. Retrieved on January 20, 2011, from http://www.evs.anl.gov/pub/doc/Water%20Mgmt%20in%20Marcellus-final-jul10.pdf.

Vejahati, F., Xu, Z., & Gupta, R. (2010). Trace elements in coal: Associations with coal and minerals and their behavior during coal utilization—a review. *Fuel, 89,* 904-911.

Vidic, R. D. (2010, March 18). *Sustainable water management for Marcellus Shale development.* Presented at Marcellus Shale natural gas stewardship: Understanding the environmental impact, Marcellus Shale Summit, Temple University, Philadelphia, PA. Retrieved July 29, 2010, from http://www.temple.edu/environment/NRDP_pics/shale/presentations_TUsummit/Vidic-Temple-2010.pdf.

Walther, J. V. (2009). *Essentials of geochemistry.* Second edition. Boston, MA: Jones and Bartlett Publishers.

Ward Jr., K. (2010, July 19). Environmentalists urge tougher water standards. *The Charleston Gazette.* Retrieved August 31, 2011, from http://sundaygazettemail.com/News/201007190845.

Warpinski, N. R., Branagan, P. T., Peterson, R. E., & Wolhart, S. L. (1998, March 15-18). *Mapping hydraulic fracture growth and geometry using microseismic events detected by a wireline retrievable accelerometer array.* Presented at the Society of Petroleum Engineers Gas Technology Symposium, Calgary, Alberta, Canada.

Warpinski, N. R., Walhart, S. L., & Wright, C. A. (2001, September 30-October 3). *Analysis and prediction of microseismicity induced by hydraulic fracturing.* Presented at the Society of Petroleum Engineers Annual Technical Conference, New Orleans, LA.

Waxman, H.A., Markey, E.J., & DeGette, D. (2011, April). Chemicals used in hydraulic fracturing. Retrieved August 31, 2011, from http://democrats.energycommerce.house.gov/sites/default/files/documents/Hydraulic%20Fracturing%20Report%204.18.11.pdf.

West Virginia Water Research Institute. (2010). *Zero discharge water management for horizontal shale gas well development: Technology status assessment.* Prepared for the US Department of Energy, National Energy Technology Laboratory, Department of Energy award no. DE-FE0001466. Morgantown, WV: West Virginia Water Research Institute, West Virginia University. Retrieved July 29, 2010, from http://prod75-inter1.netl.doe.gov/technologies/oil-gas/publications/ENVreports/FE0001466_TSA.pdf.

Williams, D.O. (2011, June 21). Fines for Garden Gulch drilling spills finally to be imposed after more than three years. *The Colorado Independent.* Retrieved August 31, 2011, from http://coloradoindependent.com/91659/fines-for-garden-gulch-drilling-spills-finally-to-be-imposed-after-more-than-three-years.

Winter, T. C., Harvey, J. W., Franke, O. L., & Alley, W. M. (1998). Ground water and surface water: A single resource. *US Geological Survey Circular, 1139,* 1-78.

Zielinski, R.A., & Budahn, J. R. Mode of occurrence and environmental mobility of oil-field radioactive material at US Geological Survey research site B, Osage-Skiatook Project, northeastern Oklahoma. *Applied Geochemistry, 22,* 2125-2137.

Ziemkiewicz, P. (2011, March 30). *Wastewater from gas development: chemical signatures in the Monongahela River Basin.* Presented at the EPA's Hydraulic Fracturing Technical Workshop 4, Washington, DC.

Zoback, M., Kitasei, S., & Copithorne, B. (2010, July). *Addressing the environmental risks from shale gas development.* Briefing paper 1. Washington, DC: Worldwatch Institute. Retrieved January 20, 2011, from http://www.worldwatch.org/files/pdf/Hydraulic%20Fracturing%20Paper.pdf.

Zorn, T. G., Seelbach, P. W., Rutherford, E. S., Wills, T. C., Cheng, S., & Wiley, M. J. (2008, November). *A regional-scale habitat suitability model to assess the effects of flow reduction on fish assemblages in Michigan streams.* Fisheries Division Research Report 2089. Lansing, MI: State of Michigan Department of Natural Resources. Retrieved January 20, 2011, from http://www.michigandnr.com/PUBLICATIONS/ PDFS/ifr/ifrlibra/Research/reports/2089/RR2089.pdf.

APPENDIX A: RESEARCH SUMMARY

TABLE A1. RESEARCH TASKS IDENTIFIED FOR WATER ACQUISITION

Water Acquisition: What are the potential impacts of large volume water withdrawals from ground and surface waters on drinking water resources?

Secondary Question	Research Tasks	Potential Product(s)	Report
How much water is used in hydraulic fracturing operations, and what are the sources of this water?	*Analysis of Existing Data* • Compile and analyze data submitted by nine hydraulic fracturing service companies for information on source water volume and quality requirements	• List of volume and water quality parameters that are important for hydraulic fracturing operations	2012
	• Compile and analyze data from nine oil and gas operators on the acquisition of source water for hydraulic fracturing operations	• Information on source, volume, and quality of water used for hydraulic fracturing operations	2012
	• Compile data on water use and hydraulic fracturing activity for the Susquehanna River Basin and Garfield County, CO	• Location-specific data on water use for hydraulic fraction	2012
	Prospective Case Studies • Document the source of the water used for hydraulic fracturing activities • Measure the quantity and quality of the water used at each case study location	• Location-specific examples of water acquisition, including data on the source, volume, and quality of the water	2014
How might water withdrawals affect short- and long-term water availability in an area with hydraulic fracturing activity?	*Analysis of Existing Data* • Compile data on water use, hydrology, and hydraulic fracturing activity for the Susquehanna River Basin and Garfield County, CO	• Maps of recent hydraulic fracturing activity and water usage in a humid region (Susquehanna River Basin) and a semi-arid region (Garfield County, CO)	2012
	• Compare control areas to areas with hydraulic fracturing activity	• Information on whether water withdrawals for hydraulic fracturing activities alter ground and surface water flows	2012
		• Assessment of impacts of hydraulic fracturing on water availability at various spatial and temporal scales	2012
	Prospective Case Studies • Compile information or water availability impacts due to water withdrawals from ground (DeSoto Parish, LA) and surface (Washington County, PA) waters	• Identification of short-term impacts on water availability from ground and surface water withdrawals associated with hydraulic fracturing activities	2014

Continued on next page

Water Acquisition: What are the potential impacts of large volume water withdrawals from ground and surface waters on drinking water resources?

Secondary Question	Research Tasks	Potential Product(s)	Report
Continued from previous page How might water withdrawals affect short- and long-term water availability in an area with hydraulic fracturing activity?	*Scenario Evaluations* • Conduct future scenario modeling of cumulative hydraulic fracturing-related water withdrawals in the Susquehanna River Basin and Garfield County, CO	• Identification of long-term water quantity impacts on drinking water resources due to cumulative water withdrawals for hydraulic fracturing	2014
What are the possible impacts of water withdrawals for hydraulic fracturing operations on local water quality?	*Analysis of Existing Data* • Compile data on water quality and hydraulic fracturing activity for the Susquehanna River Basin and Garfield County, CO • Analyze trends in water quality • Compare control areas to areas with intense hydraulic fracturing activity	• Maps of hydraulic fracturing activity and water quality for the Susquehanna River Basin and Garfield County, CO • Information on whether water withdrawals for hydraulic fracturing activities alter local water quality	2012 2012
	Prospective Case Studies • Measure local water quality before and after water withdrawals for hydraulic fracturing	• Identification of impacts on local water quality from water withdrawals for hydraulic fracturing	2014

TABLE A2. RESEARCH TASKS IDENTIFIED FOR CHEMICAL MIXING

Chemical Mixing: What are the possible impacts of surface spills on or near well pads of hydraulic fracturing fluids on drinking water resources?

Secondary Question	Research Tasks	Potential Product(s)	Report
What is currently known about the frequency, severity, and causes of spills of hydraulic fracturing fluids and additives?	*Analysis of Existing Data* • Compile information regarding surface spills obtained from nine oil and gas operators • Compile information on frequency, severity, and causes of spills of hydraulic fracturing fluids and additives from existing data sources	• Nationwide data on the frequency, severity, and causes of spills of hydraulic fracturing fluids and additives	2012
What are the identities and volumes of chemicals used in hydraulic fracturing fluids, and how might this composition vary at a given site and across the country?	*Analysis of Existing Data* • Compile information on hydraulic fracturing fluids and chemicals from publicly available data and data provided by nine hydraulic fracturing service companies • Identify factors that may alter hydraulic fracturing fluid composi'tion	• Description of types of hydraulic fracturing fluids and their frequency of use (subject to CBI rules) • List of chemicals used in hydraulic fracturing fluids, including concentrations (subject to CBI rules) • List of factors that determine and alter the composition of hydraulic fracturing fluids	2012 2012 2012
	Prospective Case Studies • Collect information on the chemical products used in the hydraulic fracturing fluids at the case study locations	• Illustrative examples of hydraulic fracturing fluids used in the Haynesville and Marcellus Shale plays	2014
What are the chemical, physical, and toxicological properties of hydraulic fracturing chemical additives?	*Analysis of Existing Data* • Search existing databases for chemical, physical, and toxicological properties • Prioritize list of chemicals based on their known properties for (1) further toxicological analysis or (2) to identify/modify existing analytical methods	• List of hydraulic fracturing chemicals with known chemical, physical, and toxicological properties • Identification of 10-20 possible indicators to track the fate and transport of hydraulic fracturing fluids based on known chemical, physical, and toxicological properties • Identification of hydraulic fracturing chemicals that may be of high concern, but have no or little existing toxicological information	2012 2012 2012

Continued on next page

101

Chemical Mixing: What are the possible impacts of surface spills on or near well pads of hydraulic fracturing fluids on drinking water resources?

Secondary Question	Research Tasks	Potential Product(s)	Report
Continued from previous page What are the chemical, physical, and toxicological properties of hydraulic fracturing chemical additives?	*Toxicological Analysis* • Identify chemicals currently undergoing ToxCast Phase II testing • Predict chemical, physical, and toxicological properties based on chemical structure for chemicals with unknown properties • Identify up to six hydraulic fracturing chemicals with unknown toxicity values for ToxCast screening and PPRTV development	• Lists of high, low, and unknown priority hydraulic fracturing chemicals based on known or predicted toxicity data	2012
		• Toxicological properties for up to six hydraulic fracturing chemicals that have no existing toxicological information and are of high concern	2014
	Laboratory Studies • Identify or modify existing analytical methods for selected hydraulic fracturing chemicals	• Analytical methods for detecting hydraulic fracturing chemicals	2012/14
If spills occur, how might hydraulic fracturing chemical additives contaminate drinking water resources?	*Analysis of Existing Data* • Review existing scientific literature on surface chemical spills with respect to hydraulic fracturing chemical additives or similar compounds	• Summary of existing research that describes the fate and transport of hydraulic fracturing chemical additives, similar compounds, or classes of compounds	2012
		• Identification of knowledge gaps for future research, if necessary	2012
	Retrospective Case Studies • Investigate hydraulic fracturing sites where surface spills of hydraulic fracturing fluids have occurred (Dunn County, ND; Bradford and Susquehanna Counties, PA)	• Identification of impacts (if any) to drinking water resources from surface spills of hydraulic fracturing fluids	2014
		• Identification of factors that led to impacts (if any) to drinking water resources resulting from the accidental release of hydraulic fracturing fluids	2014

TABLE A3. RESEARCH TASKS IDENTIFIED FOR WELL INJECTION

Well Injection: What are the possible impacts of the injection and fracturing process on drinking water resources?

Secondary Question	Research Tasks	Potential Product(s)	Report
How effective are current well construction practices at containing gases and fluids before, during, and after hydraulic fracturing?	*Analysis of Existing Data* • Compile and analyze data from nine oil and gas operators on well construction practices	• Data on the frequency and severity of well failures • Identification of contributing factors that may lead to well failures during hydraulic fracturing activities	2014 2014
	Retrospective Case Studies • Investigate the cause(s) of reported drinking water contamination—including testing well mechanical integrity—in Dunn County, ND, and Bradford and Susquehanna Counties, PA	• Identification of impacts (if any) to drinking water resources resulting from well failure or improper well construction • Data on the role of mechanical integrity in suspected cases of drinking water contamination due to hydraulic fracturing	2014 2014
	Prospective Case Studies • Conduct tests to assess well mechanical integrity before and after fracturing • Assess methods and tools used to isolate and protect drinking water resources from oil and gas resources before and during hydraulic fracturing	• Data on changes (if any) in mechanical integrity due to hydraulic fracturing • Identification of methods and tools used to isolate and protect drinking water resources from oil and gas resources before and during hydraulic fracturing	2014 2014
	Scenario Evaluations • Test scenarios involving hydraulic fracturing of inadequately or inappropriately constructed or designed wells	• Assessment of well failure scenarios during and after well injection that may lead to drinking water contamination	2012
Can subsurface migration of fluids or gases to drinking water resources occur, and what local geologic or man-made features may allow this?	*Analysis of Existing Data* • Compile and analyze information from nine oil and gas operators on data relating to the location of local geologic and man-made features and the location of hydraulically created fractures	• Information on the types of local geologic or man-made features that are searched for prior to hydraulic fracturing • Data on whether or not fractures interact with local geologic or man-made features and the frequency of occurrence	2012 2012

Continued on next page

Well Injection: What are the possible impacts of the injection and fracturing process on drinking water resources?		
Secondary Question	**Research Tasks**	**Report**
Continued from previous page Can subsurface migration of fluids or gases to drinking water resources occur, and what local geologic or man-made features may allow this?	*Retrospective Case Studies* • Investigate the cause(s) of reported drinking water contamination in an area where hydraulic fracturing is occurring within a USDW where the fractures may directly extend into an aquifer (Las Animas Co., CO)	• Identification of impacts (if any) to drinking water resources from hydraulic fracturing within a drinking water aquifer	2014
	Prospective Case Studies • Gather information on the location of known faults, fractures, and abandoned wells	• Identification of methods and tools used to determine existing faults, fractures, and abandoned wells	2014
		• Data on the potential for hydraulic fractures to interact with existing natural features	2014
	Scenario Evaluations • Test scenarios involving hydraulic fractures (1) interacting with nearby man-made features including abandoned or production wells, (2) reaching drinking water resources or permeable formations, and (3) interacting with existing faults and fractures	• Assessment of key conditions that may affect the interaction of hydraulic fractures with existing man-made and natural features	2012
	• Develop a simple model to determine the area of evaluation associated with a hydraulically fractured well	• Identification of the area of evaluation for a hydraulically fractured well	2012
How might hydraulic fracturing fluids change the fate and transport of substances in the subsurface through geochemical interactions?	*Laboratory Studies* • Identify hydraulic fracturing fluid chemical additives to be studied and relevant environmental media (e.g., soil, aquifer material, gas-bearing formation material)	• Data on the chemical composition and mineralogy of environmental media	2014
	• Characterize the chemical and mineralogical properties of the environmental media	• Data on reactions between hydraulic fracturing fluids and environmental media	2014
	• Determine the products of reactions between chosen hydraulic fracturing fluid chemical additives and relevant environmental media	• List of chemicals that may be mobilized during hydraulic fracturing activities	2014

Well Injection: *What are the possible impacts of the injection and fracturing process on drinking water resources?*

Secondary Question	Research Tasks	Potential Product(s)	Report
What are the chemical, physical, and toxicological properties of substances in the subsurface that may be released by hydraulic fracturing operations?	*Analysis of Existing Data* • Compile information from existing literature on the identity of chemicals released from the subsurface • Search existing databases for chemical, physical, and toxicological properties	• List of naturally occurring substances that are known to be mobilized during hydraulic fracturing activities and their associated chemical, physical, and toxicological properties • Identification of chemicals that may warrant further toxicological analysis or analytical method development	2012 2012
	Toxicological Analysis • Identify chemicals currently undergoing ToxCast Phase II testing • Predict chemical, physical, and toxicological properties based on chemical structure for chemicals with unknown properties (if any) • Identify up to six chemicals with unknown toxicity values for ToxCast screening and PPRTV development (if any)	• Lists of high, low, and unknown priority for naturally occurring substances based on known or predicted toxicity data • Toxicological properties for up to six naturally occurring substances that have no existing toxicological information and are of high concern	2012 2014
	Laboratory Studies • Identify or modify existing analytical methods for selected naturally occurring substances released by hydraulic fracturing	• Analytical methods for detecting selected naturally occurring substances released by hydraulic fracturing	2012/14

TABLE A4. RESEARCH TASKS IDENTIFIED FOR FLOWBACK AND PRODUCED WATER

Flowback and Produced Water:
What are the possible impacts of surface spills on or near well pads of flowback and produced water on drinking water resources?

Secondary Question	Research Tasks	Potential Product(s)	Report
What is currently known about the frequency, severity, and causes of spills of flowback and produced water?	*Analysis of Existing Data* • Compile information on frequency, severity, and causes of spills of flowback and produced waters from existing data sources	• Data on the frequency, severity, and causes of spills of flowback and produced waters	2012
What is the composition of hydraulic fracturing wastewaters, and what factors might influence this composition?	*Analysis of Existing Data* • Compile and analyze data submitted by nine hydraulic fracturing service companies for information on flowback and produced water • Compile and analyze data submitted by nine operators on the characterization of flowback and produced waters • Compile data from other sources, including existing literature and state reports	• List of chemicals found in flowback and produced water • Information on distribution (range, mean, median) of chemical concentrations • Identification of factors that may influence the composition of flowback and produced water • Identification of constituents of concern present in hydraulic fracturing wastewaters	2012 2012 2012 2012
	Prospective Case Studies • Collect time series samples of flowback and produced water at locations in the Haynesville and Marcellus shale plays	• Data on composition, variability, and quantity of flowback and produced water as a function of time	2014
What are the chemical, physical, and toxicological properties of hydraulic fracturing wastewater constituents? *Continued on next page*	*Analysis of Existing Data* • Search existing databases for chemical, physical, and toxicological properties of chemicals found in flowback and produced water • Prioritize list of chemicals based on their known properties for (1) further toxicological analysis or (2) to identify/modify existing analytical methods	• List of flowback and produced water constituents with known chemical, physical, and toxicological properties • Identification of 10-20 possible indicators to track the fate and transport of hydraulic fracturing wastewaters based on known chemical, physical, and toxicological properties • Identification of constituents that may be of high concern, but have no or little existing toxicological information	2012 2012 2012

Flowback and Produced Water:

What are the possible impacts of surface spills on or near well pads of flowback and produced water on drinking water resources?

Secondary Question	Research Tasks	Potential Product(s)	Report
Continued from previous page What are the chemical, physical, and toxicological properties of hydraulic fracturing wastewater constituents?	*Toxicological Analysis* • Predict chemical, physical, and toxicological properties based on chemical structure for chemicals with unknown properties	• Lists of high, low, and unknown-priority hydraulic fracturing chemicals based on known or predicted toxicity data	2012
	• Identify up to six hydraulic fracturing wastewater constituents with unknown toxicity values for ToxCast screening and PPRTV development	• Toxicological properties for up to six hydraulic fracturing wastewater constituents that have no existing toxicological information and are of high concern	2014
	Laboratory Studies • Identify or modify existing analytical methods for selected hydraulic fracturing wastewater constituents	• Analytical methods for detecting hydraulic fracturing wastewater constituents	2014
If spills occur, how might hydraulic fracturing wastewaters contaminate drinking water resources?	*Analysis of Existing Data* • Review existing scientific literature on surface chemical spills with respect to chemicals found in hydraulic fracturing wastewaters or similar compounds	• Summary of existing research that describes the fate and transport of chemicals in hydraulic fracturing wastewaters or similar compounds	2012
		• Identification of knowledge gaps for future research, if necessary	2012
	Retrospective Case Studies • Investigate hydraulic fracturing sites where surface spills of hydraulic fracturing wastewaters have occurred (Wise and Denton Counties, TX; Bradford and Susquehanna Counties, PA; Washington County, PA)	• Identification of impacts (if any) to drinking water resources from surface spills of hydraulic fracturing wastewaters	2014
		• Identification of factors that led to impacts (if any) to drinking water resources resulting from the accidental release of hydraulic fracturing wastewaters	2014

TABLE A5. RESEARCH TASKS IDENTIFIED FOR WASTEWATER TREATMENT AND WASTE DISPOSAL

Wastewater Treatment and Waste Disposal:

What are the possible impacts of inadequate treatment of hydraulic fracturing wastewaters on drinking water resources?

Secondary Question	Research Tasks	Potential Product(s)	Report
What are the common treatment and disposal methods for hydraulic fracturing wastewaters, and where are these methods practiced?	*Analysis of Existing Data* • Gather information from well files requested from nine well owners and operators on treatment and disposal practices	• Nationwide data on recycling, treatment, and disposal methods for hydraulic fracturing wastewaters	2012
	Prospective Case Studies • Gather information on recycling, treatment, and disposal practices in two different locations (Haynesville and Marcellus Shale)	• Information on wastewater recycling, treatment, and disposal practices at two specific locations	2014
How effective are conventional POTWs and commercial treatment systems in removing organic and inorganic contaminants of concern in hydraulic fracturing wastewaters?	*Analysis of Existing Data* • Gather existing data on the treatment efficiency and contaminant fate and transport through treatment trains applied to hydraulic fracturing wastewaters	• Collection of analytical data on the efficacy of existing treatment operations that treat hydraulic fracturing wastewaters • Identification of areas for further research	2014 2014
	Laboratory Studies • Pilot-scale studies on synthesized and actual hydraulic fracturing wastewater treatability via conventional POTW technology (e.g. settling/activated sludge processes) and commercial technologies (e.g. filtration, RO)	• Data on the fate and transport of hydraulic fracturing water contaminants through wastewater treatment processes, including partitioning in treatment residuals	2014
	Prospective Case Studies • Collect data on the efficacy of any treatment methods used in the case study	• Data on the efficacy of treatment methods used in two locations	2014

Wastewater Treatment and Waste Disposal:

What are the possible impacts of inadequate treatment of hydraulic fracturing wastewaters on drinking water resources?

Secondary Question	Research Tasks	Potential Product(s)	Report
What are the potential impacts from surface water disposal of treated hydraulic fracturing wastewater on drinking water treatment facilities?	*Laboratory Studies* • Conduct studies on the formation of brominated DBPs during treatment of hydraulic fracturing wastewaters	• Data on the formation of brominated DBPs from chlorination, chloramination, and ozonation treatments	2012/14
	• Collect discharge and stream/river samples in locations potentially impacted by hydraulic fracturing wastewater discharge	• Data on the inorganic species in hydraulic fracturing wastewater and other discharge sources that contribute similar species	2014
		• Contribution of hydraulic fracturing wastewater to stream/river contamination	2014
	Scenario Evaluation • Develop a simplified generic scenario of an idealized river with generalized inputs and receptors	• Identification of parameters that generate or mitigate drinking water exposure	2012
	• Develop watershed-specific versions of the simplified scenario using location-specific data and constraints	• Data on potential impacts in the Monongahela, Allegheny, or Susquehanna River networks	2014

TABLE A6. RESEARCH TASKS IDENTIFIED FOR ENVIRONMENTAL JUSTICE

Environmental Justice: Does hydraulic fracturing disproportionately occur in or near communities with environmental justice concerns?

Secondary Question	Research Tasks	Potential Product(s)	Report
Are large volumes of water being disproportionately withdrawn from drinking water resources that serve communities with environmental justice concerns?	*Analysis of Existing Data* • Compare data on locations of source water withdrawals to demographic information (e.g., race/ethnicity, income, and age)	• Maps showing locations of source water withdrawals and demographic data	2012
		• Identification of areas where there may be a disproportionate co-localization of large volume water withdrawals for hydraulic fracturing and communities with environmental justice concerns	2012
	Prospective Case Studies • Analyze demographic profiles of communities located near the case study locations	• Illustrative information on the types of communities where hydraulic fracturing occurs	2014
Are hydraulically fractured oil and gas wells disproportionately located near communities with environmental justice concerns?	*Analysis of Existing Data* • Compare data on locations of hydraulically fractured oil and gas wells to demographic information (e.g., race/ethnicity, income, and age)	• Maps showing locations of hydraulically fractured wells (subject to CBI rules) and demographic data	2012
		• Identification of areas where there may be a disproportionate co-localization of hydraulic fracturing well sites and communities with environmental justice concerns	2012
	Retrospective and Prospective Case Studies • Analyze demographic profiles of communities located near the case study locations	• Illustrative information on the types of communities where hydraulic fracturing occurs	2014
Is wastewater from hydraulic fracturing operations being disproportionately treated or disposed of (via POTWs or commercial treatment systems) in or near communities with environmental justice concerns?	*Analysis of Existing Data* • Compare data on locations of hydraulic fracturing wastewater disposal to demographic information (e.g., race/ethnicity, income, and age)	• Maps showing locations of wastewater disposal and demographic data	2012
		• Identification of areas where there may be a disproportionate co-localization of wastewater disposal and communities with environmental justice concerns	2012
	Prospective Case Studies • Analyze demographic profiles of communities located near the case study locations	• Illustrative information on the types of communities where hydraulic fracturing occurs	2014

APPENDIX B: STAKEHOLDER COMMENTS

In total, EPA received 5,521 comments that were submitted electronically to hydraulic.fracturing@epa.gov or mailed to EPA. This appendix provides a summary of those comments.

More than half of the electronic comments received consisted of a form letter written by Energycitizens.org[14] and sent by citizens. This letter states that "Hydraulic fracturing has been used safely and successfully for more than six decades to extract natural gas from shale and coal deposits. In this time, there have been no confirmed incidents of groundwater contamination caused by the hydraulic fracturing process." Additionally, the letter states that protecting the environment "should not lead to the creation of regulatory burdens or restrictions that have no valid scientific basis." EPA has interpreted this letter to mean that the sender supports hydraulic fracturing and does not support the need for additional study.

Table B1 provides an overall summary of the 5,521 comments received[15].

TABLE B1. SUMMARY OF STAKEHOLDER COMMENTS

Stakeholder Comments	Percentage of Comments (w/ Form Letter)	Percentage of Comments (w/o Form Letter)
Position on Study Plan		
For	18.2	63.2
Opposed	72.1	3.0
No Position	9.7	33.8
Expand Study	8.8	30.5
Limit Study	0.7	2.5
Position on Hydraulic Fracturing		
For	75.7	15.7
Opposed	11.6	40.3
No Position	12.7	44.1

Table B2 further provides the affiliations (i.e., citizens, government, industry) associated with the stakeholders, and indicates that the majority of comments EPA received came from citizens.

---

[14] Energy Citizens is financially sponsored by API, as noted at http://energycitizens.org/ec/advocacy/content-rail.aspx?ContentPage=About.

[15] Comments may be found at http://yosemite.epa.gov/sab/SABPRODUCT.NSF/81e39f4c09954fcb85256ead006be86e/d3483ab445ae61418525775900603e79!OpenDocument&TableRow=2.2#2

TABLE B2. SUMMARY OF COMMENTS ON HYDRAULIC FRACTURING AND RELATED STUDY PLAN

Category	Percentage of Comments (w/ Form Letter)	Percentage of Comments (w/o Form Letter)
Association	0.24	0.82
Business association	0.69	2.39
Citizen	23.47	81.56
Citizen (form letter Energycitizens.org)	71.22	NA
Elected official	0.18	0.63
Environmental	1.10	3.84
Federal government	0.07	0.25
Lobbying organization	0.04	0.13
Local government	0.62	2.14
Oil and gas association	0.09	0.31
Oil and gas company	0.38	1.32
Political group	0.16	0.57
Private company	0.78	2.71
Scientific organization	0.02	0.06
State government	0.13	0.44
University	0.24	0.82
Water utility	0.02	0.06
Unknown	0.56	1.95

Table B3 provides a summary of the frequent research areas requested in the stakeholder comments.

TABLE B3. FREQUENT RESEARCH AREAS REQUESTED IN STAKEHOLDER COMMENTS

Research Area	Number of Requests*
Ground water	292
Surface water	281
Air pollution	220
Water use (source of water used)	182
Flowback treatment/disposal	170
Public health	165
Ecosystem effects	160
Toxicity and chemical identification	157
Chemical fate and transport	107
Radioactivity issues	74
Seismic issues	36
Noise pollution	26

* Out of 485 total requests to expand the hydraulic fracturing study.

In addition to the frequently requested research areas, there were a variety of other comments and recommendations related to potential research areas. These comments and recommendations are listed below:

- Abandoned and undocumented wells
- Auto-immune diseases related to hydraulic fracturing chemicals
- Bioaccumulation of hydraulic fracturing chemicals in the food chain
- Biodegradable/nontoxic fracturing liquids
- Carbon footprint of entire hydraulic fracturing process
- Comparison of accident rates to coal/oil mining accident rates
- Disposal of drill cuttings
- Effects of aging on well integrity
- Effects of hydraulic fracturing on existing public and private wells
- Effects of truck/tanker traffic
- Effects on local infrastructure (e.g., roads, water treatment plants)
- Effects on tourism
- Hydraulic fracturing model
- Economic impacts on landowners
- Land farming on fracturing sludge
- Light pollution
- Long-term corrosive effects of brine and microbes on well pipes
- Natural flooding near hydraulic fracturing operations
- Radioactive proppants
- Recovery time and persistence of hydraulic fracturing chemicals in contaminated aquifers
- Recycling of flowback and produced water
- Removal of radium and other radionuclides from flowback and produced water
- Restoration of drill sites
- Review current studies of hydraulic fracturing with microseismic testing
- Sociological effects (e.g., community changes with influx of workers)
- Soil contamination at drill sites
- Volatile organic compound emissions from hydraulic fracturing operations and impoundments
- Wildlife habitat fragmentation
- Worker occupational health

APPENDIX C: DEPARTMENT OF ENERGY'S EFFORTS ON HYDRAULIC FRACTURING

DOE has invested in research on safer hydraulic fracturing techniques, including research related to well integrity, greener additives, risks from abandoned wells, possible seismic impacts, water treatment and recycling, and fugitive methane emissions.

DOE's experience includes quantifying and evaluating potential risks resulting from the production and development of shale gas resources, including multi-phase flow in wells and reservoirs, well control, casing, cementing, drilling fluids, and abandonment operations associated with drilling, completion, stimulation, and production operations. DOE also has experience in evaluating seal-integrity and wellbore-integrity characteristics in the context of the protection of groundwater.

DOE has developed a wide range of new technologies and processes, including innovations that reduce the environmental impact of exploration and production, such as greener chemicals or additives used in shale gas development, flowback water treatment processes and water filtration technologies. Data from these research activities may assist decision-makers.

DOE has developed and evaluated novel imaging technologies for areal magnetic surveys for the detection of unmarked abandoned wells, and for detecting and measuring fugitive methane emissions from exploration, production, and transportation facilities. DOE also conducts research in produced water characterization, development of shale formation fracture models, development of microseismic and isotope-based comprehensive monitoring tools, and development of integrated assessment models to predict geologic behavior during the evolution of shale gas plays. DOE's experience in engineered underground containment systems for CO_2 storage and enhanced geothermal systems also brings capabilities that are relevant to the challenges of safe shale gas production.

As part of these efforts, EPA and DOE are working together on a prospective case study located in the Marcellus Shale region that leverages DOE's capabilities in field-based monitoring of environmental signals. DOE is conducting soil gas surveys, hydraulic fracturing tracer studies, and electromagnetic induction surveys to identify possible migration of natural gas, completion fluids, or production fluids. Monitoring activities will continue throughout the development of the well pad, and during hydraulic fracturing and production of shale gas at the site. The Marcellus Test Site is undergoing a comprehensive monitoring plan, including potential impacts to drinking water resources.

More information can be found on the following websites:

- http://www.fe.doe.gov/programs/oilgas/index.html
- http://www.netl.doe.gov/technologies/oil-gas/index.html
- http://www.netl.doe.gov/kmd/Forms/Search.aspx
- http://ead.anl.gov/index.cfm
- http://www1.eere.energy.gov/geothermal/

APPENDIX D: INFORMATION REQUESTS

Request to hydraulic fracturing service companies. In September 2010, EPA issued information requests to nine hydraulic fracturing service companies to collect data that will inform this study. The requests were sent to the following companies: BJ Services, Complete Well Services, Halliburton, Key Energy Services, Patterson-UTI, RPC, Schlumberger, Superior Well Services, and Weatherford. These companies are a subset of those from which the House Committee on Energy and Commerce requested comment. Halliburton, Schlumberger, and BJ Services are the three largest companies operating in the US; the others are companies of varying size that operate in the major US shale plays. EPA sought information on the chemical composition of fluids used in the hydraulic fracturing process, data on the impacts of the chemicals on human health and the environment, standard operating procedures at hydraulic fracturing sites and the locations of sites where fracturing has been conducted. EPA sent a mandatory request to Halliburton on November 9, 2010, to compel Halliburton to provide the requested information. All companies have submitted the information.

The questions asked in the voluntary information request are stated below.

QUESTIONS

Your response to the following questions is requested within thirty (30) days of receipt of this information request:

1. Provide the name of each hydraulic fracturing fluid formulation/mixture distributed or utilized by the Company within the past five years from the date of this letter. For each formulation/mixture, provide the following information for each constituent of such product. "Constituent" includes each and every component of the product, including chemical substances, pesticides, radioactive materials and any other components.

    a. Chemical name (e.g., benzene—use IUPAC nomenclature);

    b. Chemical formula (e.g., C_6H_6);

    c. Chemical Abstract System number (e.g., 71-43-2);

    d. Material Safety Data Sheet;

    e. Concentration (e.g., ng/g or ng/L) of each constituent in each hydraulic fracturing fluid product. Indicate whether the concentration was calculated or determined analytically. This refers to the actual concentration injected during the fracturing process following mixing with source water, and the delivered concentration of the constituents to the site. Also indicate the analytical method which may be used to determine the concentration (e.g., SW-846 Method 8260, in-house SOP), and include the analytical preparation method (e.g., SW-846 Method 5035), where applicable;

    f. Identify the persons who manufactured each product and constituent and the persons

---

who sold them to the Company, including address and telephone numbers for any such persons;

g. Identify the purpose and use of each constituent in each hydraulic fracturing fluid product (e.g., solvent, gelling agent, carrier);

h. For proppants, identify the proppant, whether or not it was resin coated, and the materials used in the resin coating;

i. For the water used, identify the quantity, quality and the specifications of water needed to meet site requirements, and the rationale for the requirements;

j. Total quantities of each constituent used in hydraulic fracturing and the related quantity of water in which the chemicals were mixed to create the fracturing fluids to support calculated and/or measured composition and properties of the hydraulic fracturing fluids; and

k. Chemical and physical properties of all chemicals used, such as Henry's law coefficients, partitioning coefficients (e.g., K_{ow} $K_{OC,}$ K_d), aqueous solubility, degradation products and constants and others.

2. Provide all data and studies in the Company's possession relating to the human health and environmental impacts and effects of all products and constituents identified in Question 1.

3. For all hydraulic fracturing operations for natural gas extraction involving any of the products and constituents identified in the response to Question 1, describe the process including the following:

a. Please provide any policies, practices and procedures you employ, including any Standard Operating Procedures (SOPs) concerning hydraulic fracturing sites, for all operations including but not limited to: drilling in preparation for hydraulic fracturing including calculations or other indications for choice and composition of drilling fluids/muds; water quality characteristics needed to prepare fracturing fluid; relationships among depth, pressure, temperature, formation geology, geophysics and chemistry and fracturing fluid composition and projected volume; determination of estimated volumes of flowback and produced waters; procedures for managing flowback and produced waters; procedures to address unexpected circumstances such as loss of drilling fluid/mud, spills, leaks or any emergency conditions (e.g., blow outs), less than fully effective well completion; modeling and actual choice of fracturing conditions such as pressures, temperatures, and fracturing material choices; determination of exact concentration of constituents in hydraulic fracturing fluid formulations/mixtures; determination of dilution ratios for hydraulic fracturing fluids, and

b. Describe how fracturing fluid products and constituents are modified at a site during the fluid injection process.

    a.   Identify all sites where, and all persons to whom, the Company:

        i.   provided hydraulic fracturing fluid services that involve the use of hydraulic fracturing fluids for the year prior to the date of this letter, and

        ii.   plans to provide hydraulic fracturing fluid services that involve the use of hydraulic fracturing fluids during one year after the date of this letter.

    b.   Describe the specific hydraulic fracturing fluid services provided or to be provided for each of the sites in Question 4.a.i. and ii., including the identity of any contractor that the Company has hired or will hire to provide any portion of such services.

For each site identified in response to Question 4, please provide all information specified in the enclosed electronic spreadsheet.

Request to Oil and Gas Operators. On August 11, 2011, EPA sent letters to nine companies that own or operate oil and gas wells requesting their voluntary participation in EPA's hydraulic fracturing study. **Clayton Williams Energy, Conoco Phillips, EQT Production, Hogback Exploration, Laramie Energy II, MDS Energy, Noble Energy, Sand Ridge Operating, and Williams Production** were randomly selected from a list of operators derived from the information gathered from the September 2010 letter to hydraulic fracturing service companies. The companies were asked to provide data on well construction, design, and well operation practices for 350 oil and gas wells that were hydraulically fractured from 2009 to 2010. EPA made this request as part of its national study to examine the potential impacts of hydraulic fracturing on drinking water resources. As of October 31, 2011, all nine companies have agreed to assist EPA and are currently sending or have completed sending their information.

The wells were selected using a stratified random method and reflect diversity in both geography and size of the oil and gas operator. To identify the wells for this request, the list of operators was sort in order by those with the most wells to those with the fewest wells. EPA defined operators to be "large" if their combined number of wells accounted for the top 50 percent of wells on the list, "medium" if their combined number of wells accounted for the next 25 percent of wells on the list and "small" if their number of wells were among the last 25 percent of wells on the list. To minimize potential burden on the smallest operators, all operators with nine wells or less were removed from consideration for selection. Then, using a map from the U.S. Energy Information Administration showing all shale gas plays (Figure 3), EPA classified four different areas of the nation: East, South, Rocky Mountain (including California) and Other. To choose the nine companies that received the request, EPA randomly selected one "large" operator from each geographic area, for a total of four "large" operators, and then randomly, and without geographic consideration, selected two "medium" and three "small" operators. Once the nine companies were identified, we used a computer algorithm that balanced geographic diversity and random selection within an operator's list to select 350 wells.

The questions asked in the letters were as follows:

Your response to the following questions is requested within thirty (30) days of receipt of this information request:

For each well listed in Enclosure 5 of this letter, provide any and all of the following information:

Geologic Maps and Cross Sections

1. Prospect geologic maps of the field or area where the well is located. The map should depict, to the extent known, the general field area, including the existing production wells within the field, preferably showing surface and bottom-hole locations, names of production wells, faults within the area, locations of delineated source water protection areas, and geologic structure.
2. Geologic cross section(s) developed for the field in order to understand the geologic conditions present at the wellbore, including the directional orientation of each cross section such as north, south, east, and west.

Drilling and Completion Information

3. Daily drilling and completion records describing the day-by-day account and detail of drilling and completion activities.
4. Mud logs displaying shows of gas or oil, losses of circulation, drilling breaks, gas kicks, mud weights, and chemical additives used.
5. Caliper, density, resistivity, sonic, spontaneous potential, and gamma logs.
6. Casing tallies, including the number, grade, and weight of casing joints installed.
7. Cementing records for each casing string, which are expected to include the type of cement used, cement yield, and wait-on-cement times.
8. Cement bond logs, including the surface pressure during each logging run, and cement evaluation logs, radioactive tracer logs or temperature logs, if available.
9. Pressure testing results of installed casing.
10. Up-to-date wellbore diagram.

Water Quality, Volume, and Disposition

11. Results from any baseline water quality sampling and analyses of nearby surface or groundwater prior to drilling.
12. Results from any post-drilling and post-completion water quality sampling and analyses of nearby surface or groundwater.
13. Results from any formation water sampling and analyses, including data on composition, depth sampled, and date collected.
14. Results from chemical, biological, and radiological analyses of "flowback," including date sampled and cumulative volume of "flowback" produced since fracture stimulation.

15. Results from chemical, biological, and radiological analyses of "produced water," including date sampled and cumulative volume of "produced water" produced since fracture stimulation.
16. Volume and final disposition of "flowback."
17. Volume and final disposition of "produced water."
18. If any of the produced water or flowback fluids were recycled, provide information, including, but not limited to, recycling procedure, volume of fluid recycled, disposition of any recycling waste stream generated, and what the recycled fluids were used for.

Hydraulic Fracturing

19. Information about the acquisition of the base fluid used for fracture stimulation, including, but not limited to, its total volume, source, and quality necessary for successful stimulation. If the base fluid is not water, provide the chemical name(s) and CAS number(s) of the base fluid.
20. Estimate of fracture growth and propagation prior to hydraulic fracturing. This estimate should include modeling inputs (e.g., permeability, Young's modulus, Poisson's ratio) and outputs (e.g., fracture length, height, and width).
21. Fracture stimulation pumping schedule or plan, which would include the number, length, and location of stages; perforation cluster spacings; and the stimulation fluid to be used, including the type and respective amounts of base fluid, chemical additives and proppants planned.
22. Post-fracture stimulation report containing, but not limited to, a chart showing all pressures and rates monitored during the stimulation; depths stimulated; number of stages employed during stimulation; calculated average width, height, and half-length of fractures; and fracture stimulation fluid actually used, including the type and respective amounts of base fluid, chemical additives and proppants used.
23. Micro-seismic monitoring data associated with the well(s) listed in Enclosure 5, or conducted in a nearby well and used to set parameters for hydraulic fracturing design.

Environmental Releases

24. Spill incident reports for any fluid spill associated with this well, including spills by vendors and service companies. This information should include, but not be limited to, the volume spilled, volume recovered, disposition of any recovered volume, and the identification of any waterways or groundwater that was impacted from the spill and how this is known.

APPENDIX E: CHEMICALS IDENTIFIED IN HYDRAULIC FRACTURING FLUID AND FLOWBACK/PRODUCED WATER

NOTE: In all tables in Appendix E, the chemicals are primarily listed as identified in the cited reference. Due to varying naming conventions or errors in reporting, there may be some duplicates or inaccurate names. Some effort has been made to eliminate errors, but further evaluation will be conducted as part of the study analysis.

TABLE E1. CHEMICALS FOUND IN HYDRAULIC FRACTURING FLUIDS

Chemical Name	Use	Ref.
1-(1-naphthylmethyl)quinolinium chloride		12
1-(phenylmethyl)-ethyl pyridinium, methyl derive.	Acid corrosion inhibitor	1,6,13
1,1,1-Trifluorotoluene		7
1,1':3',1"-Terphenyl		8
1,1':4',1"-Terphenyl		8
1,1-Dichloroethylene		7
1,2,3-Propanetricarboxylic acid, 2-hydroxy-, trisodium salt, dihydrate		12,14
1,2,3-Trimethylbenzene		12, 14
1,2,4-Butanetricarboxylic acid, 2-phosphono-		12,14
1,2,4-Trimethylbenzene	Non-ionic surfactant	5,10,12,13,14
1,2-Benzisothiazolin-3-one		7,12,14
1,2-Dibromo-2,4-dicyanobutane		12,14
1,2-Ethanediaminium, N, N'-bis[2-[bis(2-hydroxyethyl)methylammonio]ethyl]-N,N'bis(2-hydroxyethyl)-N,N'-dimethyl-,tetrachloride		12
1,2-Propylene glycol		8,12,14
1,2-Propylene oxide		12
1,3,5-Triazine-1,3,5(2H,4H,6H)-triethanol		12,14
1,3,5-Trimethylbenzene		12,14
1,4-Dichlorobutane		7
1,4-Dioxane		7,14
1,6 Hexanediamine	Clay control	13
1,6-Hexanediamine		8,12
1,6-Hexanediamine dihydrochloride		12
1-[2-(2-Methoxy-1-methylethoxy)-1-methylethoxy]-2-propanol		13
1-3-Dimethyladamantane		8
1-Benzylquinolinium chloride	Corrosion inhibitor	7,12,14
1-Butanol		7,12,14
1-Decanol		12
1-Eicosene		7,14
1-Hexadecene		7,14
1-Hexanol		12
1-Methoxy-2-propanol		7,12,14
1-Methylnaphthalene		1

Table continued on next page

Table E1 continued from previous page

Chemical Name	Use	Ref.
1-Octadecanamine, N,N-dimethyl-		12
1-Octadecene		7,14
1-Octanol		12
1-Propanaminium, 3-amino-N-(carboxymethyl)-N,N-dimethyl-, N-coco acyl derivs., chlorides, sodium salts		12
1-Propanaminium, 3-amino-N-(carboxymethyl)-N,N-dimethyl-, N-coco acyl derivs., inner salts		7,12,14
1-Propanaminium, N-(3-aminopropyl)-2-hydroxy-N,N-dimethyl-3-sulfo-, N-coco acyl derivs., inner salts		7,12,14
1-Propanesulfonic acid, 2-methyl-2-[(1-oxo-2-propenyl)amino]-		7,14
1-Propanol	Crosslinker	10,12,14
1-Propene		13
1-Tetradecene		7,14
1-Tridecanol		12
1-Undecanol	Surfactant	13
2-(2-Butoxyethoxy)ethanol	Foaming agent	1
2-(2-Ethoxyethoxy)ethyl acetate		12,14
2-(Hydroxymethylamino)ethanol		12
2-(Thiocyanomethylthio)benzothiazole	Biocide	13
2,2'-(Octadecylimino)diethanol		12
2,2,2-Nitrilotriethanol		8
2,2'-[Ethane-1,2-diylbis(oxy)]diethanamine		12
2,2'-Azobis-{2-(imidazlin-2-yl)propane dihydrochloride		7,14
2,2-Dibromo-3-nitrilopropionamide	Biocide	1,6,7,9,10,12,14
2,2-Dibromopropanediamide		7,14
2,4,6-Tribromophenol		7
2,4-Dimethylphenol		4
2,4-Hexadienoic acid, potassium salt, (2E,4E)-		7,14
2,5 Dibromotoluene		7
2-[2-(2-Methoxyethoxy)ethoxy]ethanol		8
2-acrylamido-2-methylpropanesulphonic acid sodium salt polymer		12
2-acrylethyl(benzyl)dimethylammonium Chloride		7,14
2-bromo-3-nitrilopropionamide	Biocide	1,6
2-Butanone oxime		12
2-Butoxyacetic acid		8
2-Butoxyethanol	Foaming agent, breaker fluid	1,6,9,12,14
2-Butoxyethanol phosphate		8
2-Di-n-butylaminoethanol		12,14
2-Ethoxyethanol	Foaming agent	1,6
2-Ethoxyethyl acetate	Foaming agent	1
2-Ethoxynaphthalene		7,14
2-Ethyl-1-hexanol		5,12,14
2-Ethyl-2-hexenal	Defoamer	13
2-Ethylhexanol		9
2-Fluorobiphenyl		7

Table continued on next page

Table E1 continued from previous page

Chemical Name	Use	Ref.
2-Fluorophenol		7
2-Hydroxyethyl acrylate		12,14
2-Mercaptoethanol		12
2-Methoxyethanol	Foaming agent	1
2-Methoxyethyl acetate	Foaming agent	1
2-Methyl-1-propanol	Fracturing fluid	12,13,14
2-Methyl-2,4-pentanediol		12,14
2-Methyl-3(2H)-isothiazolone	Biocide	12,13
2-Methyl-3-butyn-2-ol		7,14
2-Methylnaphthalene		1
2-Methylquinoline hydrochloride		7,14
2-Monobromo-3-nitrilopropionamide	Biocide	10,12,14
2-Phosphonobutane-1,2,4-tricarboxylic acid, potassium salt		12
2-Propanol, aluminum salt		12
2-Propen-1-aminium, N,N-dimethyl-N-2-propenyl-, chloride		7,14
2-Propen-1-aminium, N,N-dimethyl-N-2-propenyl-, chloride, homopolymer		7,14
2-Propenoic acid, polymer with sodium phosphinate		7,14
2-Propenoic acid, telomer with sodium hydrogen sulfite		7,14
2-Propoxyethanol	Foaming agent	1
2-Substituted aromatic amine salt		12,14
3,5,7-Triazatricyclo(3.3.1.1(superscript 3,7))decane, 1-(3-chloro-2-propenyl)-, chloride, (Z)-		7,14
3-Bromo-1-propanol	Microbiocide	1
4-(1,1-Dimethylethyl)phenol, methyloxirane, formaldehyde polymer		7,14
4-Chloro-3-methylphenol		4
4-Dodecylbenzenesulfonic acid		7,12,14
4-Ethyloct-1-yn-3-ol	Acid inhibitor	5,12,14
4-Methyl-2-pentanol		12
4-Methyl-2-pentanone		5
4-Nitroquinoline-1-oxide		7
4-Terphenyl-d14		7
(4R)-1-methyl-4-(prop-1-en-2-yl)cyclohexene		5,12,14
5-Chloro-2-methyl-3(2H)-isothiazolone	Biocide	12,13,14
6-Methylquinoline		8
Acetaldehyde		12,14
Acetic acid	Acid treatment, buffer	5,6,9,10,12,14
Acetic acid, cobalt(2+) salt		12,14
Acetic acid, hydroxy-, reaction products with triethanolamine		14
Acetic anhydride		5,9,12,14
Acetone	Corrosion Inhibitor	5,6,12,14
Acetonitrile, 2,2',2''-nitrilotris-		12
Acetophenone		12

Table continued on next page

Table E1 continued from previous page

Chemical Name	Use	Ref.
Acetylene		9
Acetylenic alcohol		12
Acetyltriethyl citrate		12
Acrolein	Biocide	13
Acrylamide		7,12,14
Acrylamide copolymer		12
Acrylamide-sodium acrylate copolymer		7,14
Acrylamide-sodium-2-acrylamido-2-methlypropane sulfonate copolymer	Gelling agent	7,12,14
Acrylate copolymer		12
Acrylic acid/2-acrylamido-methylpropylsulfonic acid copolymer		12
Acrylic copolymer		12
Acrylic polymers		12,14
Acrylic resin		14
Acyclic hydrocarbon blend		12
Adamantane		8
Adipic acid	Linear gel polymer	6,12,14
Alcohol alkoxylate		12
Alcohols		12,14
Alcohols, C11-14-iso-, C13-rich		7,14
Alcohols, C9-C22		12
Alcohols; C12-14-secondary		12,14
Aldehyde	Corrosion inhibitor	10,12,14
Aldol		12,14
Alfa-alumina		12,14
Aliphatic acids		7,12,14
Aliphatic alcohol glycol ether		14
Aliphatic alcohol polyglycol ether		12
Aliphatic amine derivative		12
Aliphatic hydrocarbon (naphthalenesulfonic acide, sodium salt, isopropylated)	Surfactant	13
Alkaline bromide salts		12
Alkalinity		13
Alkanes, C10-14		12
Alkanes, C1-2		4
Alkanes, C12-14-iso-		14
Alkanes, C13-16-iso-		12
Alkanes, C2-3		4
Alkanes, C3-4		4
Alkanes, C4-5		4
Alkanolamine/aldehyde condensate		12
Alkenes		12
Alkenes, C>10 .alpha.-		7,12,14
Alkenes, C>8		12
Alkoxylated alcohols		12
Alkoxylated amines		12
Alkoxylated phenol formaldehyde resin		12,14

Table continued on next page

Table E1 continued from previous page

Chemical Name	Use	Ref.
Alkyaryl sulfonate		12
Alkyl alkoxylate		12,14
Alkyl amine		12
Alkyl amine blend in a metal salt solution		12,14
Alkyl aryl amine sulfonate		12
Alkyl aryl polyethoxy ethanol		7,14
Alkyl esters		12,14
Alkyl hexanol		12,14
Alkyl ortho phosphate ester		12
Alkyl phosphate ester		12
Alkyl quaternary ammonium chlorides		12
Alkyl* dimethyl benzyl ammonium chloride *(61% C12, 23% C14, 11% C16, 2.5% C18 2.5% C10 and trace of C8)	Corrosion inhibitor	7
Alkylaryl sulfonate		7,12,14
Alkylaryl sulphonic acid		12
Alkylated quaternary chloride		12,14
Alkylbenzenesulfonate, linear	Foaming agent	5,6,12
Alkylbenzenesulfonic acid		9,12,14
Alkylethoammonium sulfates		12
Alkylphenol ethoxylates		12
Almandite and pyrope garnet		12,14
Alpha-C11-15-sec-alkyl-omega-hydroxypoly(oxy-1,2-ethanediyl)		12
Alpha-Terpineol		8
Alumina	Proppant	12,13,14
Aluminium chloride		7,12,14
Aluminum	Crosslinker	4,6,12,14
Aluminum oxide		12,14
Aluminum oxide silicate		12
Aluminum silicate	Proppant	13,14
Aluminum sulfate		12,14
Amides, coco, N-[3-(dimethylamino)propyl]		12,14
Amides, coco, N-[3-(dimethylamino)propyl], alkylation products with chloroacetic acid, sodium salts		12
Amides, coco, N-[3-(dimethylamino)propyl], N-oxides		7,12,14
Amides, tall-oil fatty, N,N-bis(hydroxyethyl)		7,14
Amides, tallow, n-[3-(dimethylamino)propyl],n-oxides		12
Amidoamine		12
Amine		12,14
Amine bisulfite		12
Amine oxides		12
Amine phosphonate		12
Amine salt		12
Amines, C14-18; C16-18-unsaturated, alkyl, ethoxylated		12
Amines, C8-18 and C18-unsatd. alkyl	Foaming agent	5
Amines, coco alkyl, acetate		12
Amines, coco alkyl, ethoxylated		14

Table continued on next page

Table E1 continued from previous page

Chemical Name	Use	Ref.
Amines, polyethylenepoly-, ethoxylated, phosphonomethylated		12
Amines, tallow alkyl, ethoxylated, acetates (salts)		12,14
Amino compounds		12
Amino methylene phosphonic acid salt		12
Aminotrimethylene phosphonic acid		12
Ammonia		9,11,12,14
Ammonium acetate	Buffer	5,10,12,14
Ammonium alcohol ether sulfate		7,12,14
Ammonium bifluoride		9
Ammonium bisulfite	Oxygen scavenger	3,9,12,14
Ammonium C6-C10 alcohol ethoxysulfate		12
Ammonium C8-C10 alkyl ether sulfate		12
Ammonium chloride	Crosslinker	1,6,10,12,14
Ammonium citrate		7,14
Ammonium fluoride		12,14
Ammonium hydrogen carbonate		12,14
Ammonium hydrogen difluoride		12,14
Ammonium hydrogen phosphonate		14
Ammonium hydroxide		7,12,14
Ammonium nitrate		7,12,14
Ammonium persulfate	Breaker fluid	1,6,9
Ammonium salt		12,14
Ammonium salt of ethoxylated alcohol sulfate		12,14
Ammonium sulfate	Breaker fluid	5,6,12,14
Amorphous silica		9,12,14
Anionic copolymer		12,14
Anionic polyacrylamide		12,14
Anionic polyacrylamide copolymer	Friction reducer	5,6,12
Anionic polymer		12,14
Anionic polymer in solution		12
Anionic surfactants	Friction reducer	5,6
Anionic water-soluble polymer		12
Anthracene		4
Antifoulant		12
Antimonate salt		12,14
Antimony		7
Antimony pentoxide		12
Antimony potassium oxide		12,14
Antimony trichloride		12
Aromatic alcohol glycol ether		12
Aromatic aldehyde		12
Aromatic hydrocarbons		13,14
Aromatic ketones		12,14
Aromatic polyglycol ether		12
Aromatics		1
Arsenic		4
Arsenic compounds		14

Table continued on next page

Table E1 continued from previous page

Chemical Name	Use	Ref.
Ashes, residues		14
Atrazine		8
Attapulgite	Gelling agent	13
Barium		4
Barium sulfate		5,12,14
Bauxite	Proppant	12,13,14
Bentazone		8
Bentone clay		14
Bentonite	Fluid additives	5,6,12,14
Bentonite, benzyl(hydrogenated tallow alkyl) dimethylammonium stearate complex		14
Benzalkonium chloride		14
Benzene	Gelling agent	1,12,14
Benzene, 1,1'-oxybis-, tetrapropylene derivs., sulfonated, sodium salts		14
Benzene, C10-16-alkyl derivs.		12
Benzenesulfonic acid, (1-methylethyl)-, ammonium salt		7,14
Benzenesulfonic acid, C10-16-alkyl derivs.		12,14
Benzenesulfonic acid, C10-16-alkyl derivs., potassium salts		12,14
Benzo(a)pyrene		4
Benzoic acid		9,12,14
Benzyl chloride		12
Benzyl-dimethyl-(2-prop-2-enoyloxyethyl)ammonium chloride		8
Benzylsuccinic acid		8
Beryllium		11
Bicarbonate		7
Bicine		12
Biocide component		12
Bis(1-methylethyl)naphthalenesulfonic acid, cyclohexylamine salt		12
Bis(2-methoxyethyl) ether	Foaming Agent	1
Bishexamethylenetriamine penta methylene phosphonic acid		12
Bisphenol A		8
Bisphenol A/Epichlorohydrin resin		12,14
Bisphenol A/Novolac epoxy resin		12,14
Blast furnace slag	Viscosifier	13,14
Borate salts	Crosslinker	3,12,14
Borax	Crosslinker	1,6,12,14
Boric acid	Crosslinker	1,6,9,12,14
Boric acid, potassium salt		12,14
Boric acid, sodium salt		9,12
Boric oxide		7,12,14
Boron		4
Boron sodium oxide		12,14
Boron sodium oxide tetrahydrate		12,14

Table continued on next page

Table E1 continued from previous page

Chemical Name	Use	Ref.
Bromide (-1)		7
Bromodichloromethane		7
Bromoform		7
Bronopol	Microbiocide	5,6,12,14
Butane		5
Butanedioic acid, sulfo-, 1,4-bis(1,3-dimethylbutyl) ester, sodium salt		12
Butyl glycidyl ether		12,14
Butyl lactate		12,14
C.I. Pigment orange 5		14
C10-C16 ethoxylated alcohol	Surfactant	12,13,14
C-11 to C-14 n-alkanes, mixed		12
C12-14-tert-alkyl ethoxylated amines		7,14
Cadmium		4
Cadmium compounds		13,14
Calcium		4
Calcium bromide		14
Calcium carbonate		12,14
Calcium chloride		7,9,12,14
Calcium dichloride dihydrate		12,14
Calcium fluoride		12
Calcium hydroxide	pH control	12,13,14
Calcium hypochlorite		12,14
Calcium oxide	Proppant	9,12,13,14
Calcium peroxide		12
Calcium sulfate	Gellant	13,14
Carbohydrates		5,12,14
Carbon		14
Carbon black	Resin	13,14
Carbon dioxide	Foaming agent	5,6,12,14
Carbonate alkalinity		7
Carbonic acid calcium salt (1:1)	pH control	12,13
Carbonic acid, dipotassium salt		12,14
Carboxymethyl cellulose		8
Carboxymethyl guar gum, sodium salt		12
Carboxymethyl hydroxypropyl guar		9,12,14
Carboxymethylguar	Linear gel polymer	6
Carboxymethylhydroxypropylguar	Linear gel polymer	6
Cationic polymer	Friction reducer	5,6
Caustic soda		13,14
Caustic soda beads		13,14
Cellophane		12,14
Cellulase enzyme		12
Cellulose		7,12,14
Cellulose derivative		12,14
Ceramic		13,14
Cetyl trimethyl ammonium bromide		12
CFR-3		14

Table continued on next page

Table E1 continued from previous page

Chemical Name	Use	Ref.
Chloride		4
Chloride (-1)		14
Chlorine	Lubricant	13
Chlorine dioxide		7,12,14
Chlorobenzene		4
Chlorodibromomethane		7
Chloromethane		7
Chlorous ion solution		12
Choline chloride		9,12,14
Chromates		12,14
Chromium	Crosslinker	11
Chromium (III) acetate		12
Chromium (III), insoluble salts		6
Chromium (VI)		6
Chromium acetate, basic		13
Cinnamaldehyde (3-phenyl-2-propenal)		9,12,14
Citric acid	Iron control	3,9,12,14
Citrus terpenes		7,12,14
Coal, granular		12,14
Cobalt		7
Coco-betaine		7,14
Coconut oil acid/diethanolamine condensate (2:1)		12
Collagen (gelatin)		12,14
Common White		14
Complex alkylaryl polyo-ester		12
Complex aluminum salt		12
Complex organometallic salt		12
Complex polyamine salt		9
Complex substituted keto-amine		12
Complex substituted keto-amine hydrochloride		12
Copolymer of acrylamide and sodium acrylate		12,14
Copper		5,12
Copper compounds	Breaker fluid	1,6
Copper sulfate		7,12,14
Copper(I) iodide	Breaker fluid	5,6,12,14
Copper(II) chloride		7,12,14
Coric oxide		14
Corn sugar gum	Corrosion inhibitor	12,13,14
Corundum		14
Cottonseed flour		13,14
Cremophor(R) EL		7,12,14
Crissanol A-55		7,14
Cristobalite		12,14
Crotonaldehyde		12,14
Crystalline silica, tridymite		12,14
Cumene		7,12,14
Cupric chloride dihydrate		7,9,12
Cuprous chloride		12,14

Table continued on next page

Table E1 continued from previous page

Chemical Name	Use	Ref.
Cured acrylic resin		12,14
Cured resin		9,12,14
Cured silicone rubber-polydimethylsiloxane		12
Cured urethane resin		12,14
Cyanide		11
Cyanide, free		7
Cyclic alkanes		12
Cyclohexane		9,12
Cyclohexanone		12,14
D-(-)-Lactic acid		12,14
Dapsone		12,14
Dazomet	Biocide	9,12,13,14
Decyldimethyl amine		7,14
D-Glucitol		7,12,14
D-Gluconic acid		12
D-Glucose		12
D-Limonene		5,7,9
Di(2-ethylhexyl) phthalate		7,12
Diatomaceous earth, calcined		12
Diatomaceus earth	Proppant	13,14
Dibromoacetonitrile		7,12,14
Dibutyl phthalate		4
Dicalcium silicate		12,14
Dicarboxylic acid		12
Didecyl dimethyl ammonium chloride	Biocide	12,13
Diesel		1,6,12
Diethanolamine	Foaming agent	1,6,12,14
Diethylbenzene		7,12,14
Diethylene glycol		5,9,12,14
Diethylene glycol monobutyl ether		8.
Diethylene glycol monoethyl ether	Foaming agent	1
Diethylene glycol monomethyl ether	Foaming agent	1,12,14
Diethylenetriamine	Activator	10,12,14
Diisopropylnaphthalene		7,14
Diisopropylnaphthalenesulfonic acid		7,12,14
Dimethyl glutarate		12,14
Dimethyl silicone		12,14
Dinonylphenyl polyoxyethylene		14
Dipotassium monohydrogen phosphate		5
Dipropylene glycol		7,12,14
Di-secondary-butylphenol		12
Disodium dodecyl(sulphonatophenoxy)benzenesulphonate		12
Disodium ethylenediaminediacetate		12
Disodium ethylenediaminetetraacetate dihydrate		12
Dispersing agent		12
Distillates, petroleum, catalytic reformer fractionator residue, low-boiling		12

Table continued on next page

Table E1 continued from previous page

Chemical Name	Use	Ref.
Distillates, petroleum, hydrodesulfurized light catalytic cracked		12
Distillates, petroleum, hydrodesulfurized middle		12
Distillates, petroleum, hydrotreated heavy naphthenic		5,12,14
Distillates, petroleum, hydrotreated heavy paraffinic		12,14
Distillates, petroleum, hydrotreated light	Friction reducer	5,9,10,12,14
Distillates, petroleum, hydrotreated light naphthenic		12
Distillates, petroleum, hydrotreated middle		12
Distillates, petroleum, light catalytic cracked		12
Distillates, petroleum, solvent-dewaxed heavy paraffinic		12,14
Distillates, petroleum, solvent-refined heavy naphthenic		12
Distillates, petroleum, steam-cracked		12
Distillates, petroleum, straight-run middle		12,14
Distillates, petroleum, sweetened middle		12,14
Ditallow alkyl ethoxylated amines		7,14
Docusate sodium		12
Dodecyl alcohol ammonium sulfate		12
Dodecylbenzene		7,14
Dodecylbenzene sulfonic acid salts		12,14
Dodecylbenzenesulfonate isopropanolamine		7,12,14
Dodecylbenzene sulfonic acid, monoethanolamine salt		12
Dodecylbenzene sulphonic acid, morpholine salt		12,14
Econolite Additive		14
Edifas B	Fluid additives	5,14
EDTA copper chelate	Breaker fluid, activator	5,6,10,12,14
Endo- 1,4-beta-mannanase, or Hemicellulase		14
EO-C7-9-iso; C8 rich alcohols		14
EO-C9-11-iso; C10 rich alcohols		12,14
Epichlorohydrin		12,14
Epoxy resin		12
Erucic amidopropyl dimethyl detaine		7,12,14
Essential oils		12
Ester salt	Foaming agent	1
Ethanaminium, N,N,N-trimethyl-2-[(1-oxo-2-propenyl)oxy]-, chloride		14
Ethanaminium, N,N,N-trimethyl-2-[(1-oxo-2-propenyl)oxy]-,chloride, polymer with 2-propenamide		12,14
Ethane		5
Ethanol	Foaming agent, non-ionic surfactant	1,6,10,12,14
Ethanol, 2,2'-iminobis-, N-coco alkyl derivs., N-oxides		12
Ethanol, 2,2'-iminobis-, N-tallow alkyl derivs.		12
Ethanol, 2-[2-[2-(tridecyloxy)ethoxy]ethoxy]-, hydrogen sulfate, sodium salt		12
Ethanolamine	Crosslinker	1,6,12,14
Ethoxylated 4-nonylphenol		13

Table continued on next page

Table E1 continued from previous page

Chemical Name	Use	Ref.
Ethoxylated alcohol/ester mixture		14
Ethoxylated alcohols[16]		5,9,12,13,14
Ethoxylated alkyl amines		12,14
Ethoxylated amine		12,14
Ethoxylated fatty acid ester		12,14
Ethoxylated fatty acid, coco		14
Ethoxylated fatty acid, coco, reaction product with ethanolamine		14
Ethoxylated nonionic surfactant		12
Ethoxylated nonylphenol		8,12,14
Ethoxylated propoxylated C12-14 alcohols		12,14
Ethoxylated sorbitan trioleate		7,14
Ethoxylated sorbitol esters		12,14
Ethoxylated undecyl alcohol		12
Ethoxylated, propoxylated trimethylolpropane		7,14
Ethylacetate		9,12,14
Ethylacetoacetate		12
Ethyllactate		7,14
Ethylbenzene	Gelling Agent	1,9,12,14
Ethylcellulose	Fluid Additives	13
Ethylene glycol	Crosslinker/ Breaker Fluids/ Scale Inhibitor	1,6,9,12,14
Ethylene glycol diethyl ether	Foaming Agent	1
Ethylene glycol dimethyl ether	Foaming Agent	1
Ethylene oxide		7,12,14
Ethylene oxide-nonylphenol polymer		12
Ethylenediaminetetraacetic acid		12,14
Ethylenediaminetetraacetic acid tetrasodlum salt hydrate		7,12,14
Ethylenediaminetetraacetic acid, diammonium copper salt		14
Ethylene-vinyl acetate copolymer		12
Ethylhexanol		14
Fatty acid ester		12
Fatty acid, tall oil, hexa esters with sorbitol, ethoxylated		12,14
Fatty acids		12
Fatty acids, tall oil reaction products w/acetophenone, formaldehyde & thiourea		14
Fatty acids, tall-oil		7,12,14
Fatty acids, tall-oil, reaction products with diethylenetriamine		12
Fatty acids, tallow, sodium salts		7,14
Fatty alcohol alkoxylate		12,14
Fatty alkyl amine salt		12

Table continued on next page

---

[16] Multiple categories of ethoxylated alcohols were listed in various references. Due to different naming conventions, there is some uncertainty as to whether some are duplicates or some incorrect. Therefore, "ethoxylated alcohols" is included here as a single item with further evaluation to follow.

Table E1 continued from previous page

Chemical Name	Use	Ref.
Fatty amine carboxylates		12
Fatty quaternary ammonium chloride		12
FD & C blue no. 1		12
Ferric chloride		7,12,14
Ferric sulfate		12,14
Fluorene		1
Fluoride		7
Fluoroaliphatic polymeric esters		12,14
Formaldehyde polymer		12
Formaldehyde, polymer with 4-(1,1-dimethyl)phenol, methyloxirane and oxirane		12
Formaldehyde, polymer with 4-nonylphenol and oxirane		12
Formaldehyde, polymer with ammonia and phenol		12
Formaldehyde, polymers with branched 4-nonylphenol, ethylene oxide and propylene oxide		14
Formalin		7,12,14
Formamide		7,12,14
Formic acid	Acid Treatment	1,6,9,12,14
Formic acid, potassium salt		7,12,14
Fuel oil, no. 2		12,14
Fuller's earth	Gelling agent	13.
Fumaric acid	Water gelling agent/ linear gel polymer	1,6,12,14
Furfural		12,14
Furfuryl alcohol		12,14
Galactomannan	Gelling agent	13
Gas oils, petroleum, straight-run		12
Gilsonite	Viscosifier	12,14
Glass fiber		7,12,14
Gluconic acid		9
Glutaraldehyde	Biocide	3,9,12,14
Glycerin, natural	Crosslinker	7,10,12,14
Glycine, N-(carboxymethyl)-N-(2-hydroxyethyl)-, disodium salt		12
Glycine, N,N'-1,2-ethanediylbis[N-(carboxymethyl)-, disodium salt		7,12,14
Glycine, N,N-bis(carboxymethyl)-, trisodium salt		7,12,14
Glycine, N-[2-[bis(carboxymethyl)amino]ethyl]-N-(2-hydroxyethyl)-, trisodium salt		12
Glycol ethers		9,12
Glycolic acid		7,12,14
Glycolic acid sodium salt		7,12,14
Glyoxal		12
Glyoxylic acid		12
Graphite	Fluid additives	13
Guar gum		9,12,14
Guar gum derivative		12

Table continued on next page

Table E1 continued from previous page

Chemical Name	Use	Ref.
Gypsum		13,14
Haloalkyl heteropolycycle salt		12
Heavy aromatic distillate		12
Heavy aromatic petroleum naphtha		13,14
Hematite		12,14
Hemicellulase		5,12,14
Heptane		5,12
Heptene, hydroformylation products, high-boiling		12
Hexane		5
Hexanes		12
Hydrated aluminum silicate		12,14
Hydrocarbons		12
Hydrocarbons, terpene processing by-products		7,12,14
Hydrochloric acid	Acid treatment, solvent	1,6,9,10,12,14
Hydrogen fluoride (Hydrofluoric acid)	Acid treatment	12
Hydrogen peroxide		7,12,14
Hydrogen sulfide		7,12
Hydrotreated and hydrocracked base oil		12
Hydrotreated heavy naphthalene		5
Hydrotreated light distillate		14
Hydrotreated light petroleum distillate		14
Hydroxyacetic acid ammonium salt		7,14
Hydroxycellulose	Linear gel polymer	6
Hydroxyethylcellulose	Gel	3,12,14
Hydroxylamine hydrochloride		7,12,14
Hydroxyproplyguar	Linear gel polymer	6
Hydroxypropyl cellulose		8
Hydroxypropyl guar gum	Linear gel delivery, water gelling agent	1,6,10,12,14
Hydroxysultaine		12
Igepal CO-210		7,12,14
Inner salt of alkyl amines		12,14
Inorganic borate		12,14
Inorganic particulate		12,14
Inorganic salt		12
Instant coffee purchased off the shelf		12
Inulin, carboxymethyl ether, sodium salt		12
Iron	Emulsifier/surfactant	13
Iron oxide	Proppant	12,13,14
Iron(II) sulfate heptahydrate		7,12,14
Iso-alkanes/n-alkanes		12,14
Isoascorbic acid		7,12,14
Isomeric aromatic ammonium salt		7,12,14
Isooctanol		5,12,14
Isooctyl alcohol		12
Isopentyl alcohol		12

Table continued on next page

Table E1 continued from previous page

Chemical Name	Use	Ref.
Isopropanol	Foaming agent/ surfactant, acid corrosion inhibitor	1,6,9,12,14
Isopropylamine		12
Isoquinoline, reaction products with benzyl chloride and quinoline		14
Isotridecanol, ethoxylated		7,12,14
Kerosine, petroleum, hydrodesulfurized		7,12,14
Kyanite	Proppant	12,13,14
Lactic acid		12
Lactose		7,14
Latex 2000		13,14
L-Dilactide		12,14
Lead		4,12
Lead compounds		14
Lignite	Fluid additives	13
Lime		14
Lithium		7
L-Lactic acid		12
Low toxicity base oils		12
Lubra-Beads coarse		14
Maghemite		12,14
Magnesium		4
Magnesium aluminum silicate	Gellant	13
Magnesium carbonate		12
Magnesium chloride	Biocide	12,13
Magnesium chloride hexahydrate		14
Magnesium hydroxide		12
Magnesium iron silicate		12,14
Magnesium nitrate	Biocide	12,13,14
Magnesium oxide		12,14
Magnesium peroxide		12
Magnesium phosphide		12
Magnesium silicate		12,14
Magnetite		12,14
Manganese		4
Mercury		11
Metal salt		12
Metal salt solution		12
Methanamine, N,N-dimethyl-, hydrochloride		5,12,14
Methane		5
Methanol	Acid corrosion inhibitor	1,6,9,10,12,14
Methenamine		12,14
Methyl bromide		7
Methyl ethyl ketone		4
Methyl salicylate		9
Methyl tert-butyl ether	Gelling agent	1
Methyl vinyl ketone		12

Table continued on next page

Table E1 continued from previous page

Chemical Name	Use	Ref.
Methylcyclohexane		12
Methylene bis(thiocyanate)	Biocide	13
Methyloxirane polymer with oxirane, mono (nonylphenol) ether, branched		14
Mica	Fluid additives	5,6,12,14
Microbond expanding additive		14
Mineral		12,14
Mineral filler		12
Mineral oil	Friction reducer	3,14
Mixed titanium ortho ester complexes		12
Modified lignosulfonate		14
Modified alkane		12,14
Modified cycloaliphatic amine adduct		12,14
Modified lignosulfonate		12
Modified polysaccharide or pregelatinized cornstarch or starch		8
Molybdenum		7
Monoethanolamine		14
Monoethanolamine borate		12,14
Morpholine		12,14
Muconic acid		8
Mullite		12,14
N,N,N-Trimethyl-2[1-oxo-2-propenyl]oxy ethanaminimum chloride		7,14
N,N,N-Trimethyloctadecan-1-aminium chloride		12
N,N'-Dibutylthiourea		12
N,N-Dimethyl formamide	Breaker	3,14
N,N-Dimethyl-1-octadecanamine-HCl		12
N,N-Dimethyldecylamine oxide		7,12,14
N,N-Dimethyldodecylamine-N-oxide		8
N,N-Dimethylformamide		5,12,14
N,N-Dimethyl-methanamine-n-oxide		7,14
N,N-Dimethyl-N-[2-[(1-oxo-2-propenyl)oxy]ethyl]-benzenemethanaminium chloride		7,14
N,N-Dimethyloctadecylamine hydrochloride		12
N,N'-Methylenebisacrylamide		12,14
n-Alkanes,C10-C18		4
n-Alkanes,C18-C70		4
n-Alkanes,C5-C8		4
n-Butanol		9
Naphtha, petroleum, heavy catalytic reformed		5,12,14
Naphtha, petroleum, hydrotreated heavy		7,12,14
Naphthalene	Gelling agent, non-ionic surfactant	1,9,10,12,14
Naphthalene derivatives		12
Naphthalenesulphonic acid, bis (1-methylethyl)-methyl derivatives		12
Naphthenic acid ethoxylate		14

Table continued on next page

Table E1 continued from previous page

Chemical Name	Use	Ref.
Navy fuels JP-5		7,12,14
Nickel		4
Nickel sulfate	Corrosion inhibitor	13
Nickel(II) sulfate hexahydrate		12
Nitrazepam		8
Nitrilotriacetamide	scale inhibiter	9,12
Nitrilotriacetic acid		12,14
Nitrilotriacetic acid trisodium monohydrate		12
Nitrobenzene		8
Nitrobenzene-d5		7
Nitrogen, liquid	Foaming agent	5,6,12,14
N-Lauryl-2-pyrrolidone		12
N-Methyl-2-pyrrolidone		12,14
N-Methyldiethanolamine		8
N-Oleyl diethanolamide		12
Nonane, all isomers		12
Non-hazardous salt		12
Nonionic surfactant		12
Nonylphenol (mixed)		12
Nonylphenol ethoxylate		8,12,14
Nonylphenol, ethoxylated and sulfated		12
N-Propyl zirconate		12
N-Tallowalkyltrimethylenediamines		12,14
Nuisance particulates		12
Nylon fibers		12,14
Oil and grease		4
Oil of wintergreen		12,14
Oils, pine		12,14
Olefinic sulfonate		12
Olefins		12
Organic acid salt		12,14
Organic acids		12
Organic phosphonate		12
Organic phosphonate salts		12
Organic phosphonic acid salts		12
Organic salt		12,14
Organic sulfur compound		12
Organic surfactants		12
Organic titanate		12,14
Organo-metallic ammonium complex		12
Organophilic clays		7,12,14
O-Terphenyl		7,14
Other inorganic compounds		12
Oxirane, methyl-, polymer with oxirane, mono-C10-16-alkyl ethers, phosphates		12
Oxiranemethanaminium, N,N,N-trimethyl-, chloride, homopolymer		7,14
Oxyalkylated alcohol		12,14

Table continued on next page

Table E1 continued from previous page

Chemical Name	Use	Ref.
Oxyalkylated alkyl alcohol		12
Oxyalkylated alkylphenol		7,12,14
Oxyalkylated fatty acid		12
Oxyalkylated phenol		12
Oxyalkylated polyamine		12
Oxylated alcohol		5,12,14
P/F resin		14
Paraffin waxes and hydrocarbon waxes		12
Paraffinic naphthenic solvent		12
Paraffinic solvent		12,14
Paraffins		12
Pentaerythritol		8
Pentane		5
Perlite		14
Peroxydisulfuric acid, diammonium salt	Breaker fluid	1,6,12,14
Petroleum		12
Petroleum distillates		12,14
Petroleum gas oils		12
Petroleum hydrocarbons		7
Phenanthrene	Biocide	1,6
Phenol		4,12,14
Phenolic resin	Proppant	9,12,13,14
Phosphate ester		12,14
Phosphate esters of alkyl phenyl ethoxylate		12
Phosphine		12,14
Phosphonic acid		12
Phosphonic acid (dimethlamino(methylene))		12
Phosphonic acid, (1-hydroxyethylidene)bis-, tetrasodium salt		12,14
Phosphonic acid, [[(phosphonomethyl)imino]bis[2,1-ethanediylnitrilobis(methylene)]]tetrakis-	Scale inhibitor	12,13
Phosphonic acid, [[(phosphonomethyl)imino]bis[2,1-ethanediylnitrilobis(methylene)]]tetrakis-, sodium salt		7,14
Phosphonic acid, [nitrilotris(methylene)]tris-, pentasodium salt		12
[[(Phosphonomethyl)imino]bis[2,1-ethanediylnitrilobis(methylene)]]tetrakis phosphonic acid ammonium salt		7,14
Phosphoric acid ammonium salt		12
Phosphoric acid Divosan X-Tend formulation		12
Phosphoric acid, aluminium sodium salt	Fluid additives	12,13
Phosphoric acid, diammonium salt	Corrosion inhibitor	13
Phosphoric acid, mixed decyl and Et and octyl esters		12
Phosphoric acid, monoammonium salt		14
Phosphorous acid		12
Phosphorus		7
Phthalic anhydride		12
Plasticizer		12

Table continued on next page

Table E1 continued from previous page

Chemical Name	Use	Ref.
Pluronic F-127		12,14
Poly (acrylamide-co-acrylic acid), partial sodium salt		14
Poly(oxy-1,2-ethanediyl), .alpha.-(nonylphenyl)-omega-hydroxy-, phosphate		12,14
Poly(oxy-1,2-ethanediyl), .alpha.-(octylphenyl)-omega-hydroxy-, branched		12
Poly(oxy-1,2-ethanediyl), alpha,alpha'-[[(9Z)-9-octadecenylimino]di-2,1-ethanediyl]bis[.omega.-hydroxy-		12,14
Poly(oxy-1,2-ethanediyl), alpha-sulfo-.omega.-hydroxy-, C12-14-alkyl ethers, sodium salts		12,14
Poly(oxy-1,2-ethanediyl), alpha-hydro-omega-hydroxy		12
Poly(oxy-1,2-ethanediyl), alpha-sulfo-omega-(hexyloxy)-ammonium salt		12,14
Poly(oxy-1,2-ethanediyl), alpha-tridecyl-omega-hydroxy-		12,14
Poly-(oxy-1,2-ethanediyl)-alpha-undecyl-omega-hydroxy		12,14
Poly(oxy-1,2-ethanediyl)-nonylphenyl-hydroxy	Acid corrosion inhibitor, non-ionic surfactant	7,12,13,14
Poly(sodium-p-styrenesulfonate)		12
Poly(vinyl alcohol)		12
Poly[imino(1,6-dioxo-1,6-hexanediyl)imino-1,6-hexanediyl]	Resin	13
Polyacrylamide	Friction reducer	3,6,12,13,14
Polyacrylamides		12
Polyacrylate		12,14
Polyamine		12,14
Polyamine polymer		14
Polyanionic cellulose		12
Polyaromatic hydrocarbons	Gelling agent/ bactericides	1,6,13
Polycyclic organic matter	Gelling agent/ bactericides	1,6,13
Polyethene glycol oleate ester		7,14
Polyetheramine		12
Polyethoxylated alkanol		7,14
Polyethylene glycol		5,9,12,14
Polyethylene glycol ester with tall oil fatty acid		12
Polyethylene glycol mono(1,1,3,3-tetramethylbutyl)phenyl ether		7,12,14
Polyethylene glycol monobutyl ether		12,14
Polyethylene glycol nonylphenyl ether		7,12,14
Polyethylene glycol tridecyl ether phosphate		12
Polyethylene polyammonium salt		12
Polyethyleneimine		14
Polyglycol ether	Foaming agent	1,6,13

Table continued on next page

Table E1 continued from previous page

Chemical Name	Use	Ref.
Polyhexamethylene adipamide	Resin	13
Polylactide resin		12,14
Polymer		14
Polymeric hydrocarbons		14
Polyoxyalkylenes		9,12
Polyoxylated fatty amine salt		7,12,14
Polyphosphoric acids, esters with triethanolamine, sodium salts		12
Polyphosphoric acids, sodium salts		12,14
Polypropylene glycol	Lubricant	12,13
Polysaccharide		9,12,14
Polysaccharide blend		14
Polysorbate 60		14
Polysorbate 80		7,14
Polyvinyl alcohol	Fluid additives	12,13,14
Polyvinyl alcohol/polyvinylacetate copolymer		12
Portland cement clinker		14
Potassium		7
Potassium acetate		7,12,14
Potassium aluminum silicate		5
Potassium borate		7,14
Potassium carbonate	pH control	3,10,13
Potassium chloride	Brine carrier fluid	1,6,9,12,13,14
Potassium hydroxide	Crosslinker	1,6,12,13,14
Potassium iodide		12,14
Potassium metaborate		5,12,14
Potassium oxide		12
Potassium pentaborate		12
Potassium persulfate	Fluid additives	12,13
Propane		5
Propanimidamide, 2,2''-azobis[2-methyl-, dihydrochloride		12,14
Propanol, 1(or 2)-(2-methoxymethylethoxy)-		8,12,14
Propargyl alcohol	Acid corrosion inhibitor	1,6,9,12,13,14
Propylene carbonate		12
Propylene glycol		14
Propylene pentamer		12
p-Xylene		12,14
Pyridine, alkyl derivs.		12
Pyridinium, 1-(phenylmethyl)-, Et Me derivs., chlorides	Acid corrosion inhibitor, corrosion inhibitor	1,6,12,13,14
Pyrogenic colloidal silica		12,14
Quartz	Proppant	5,6,12,13,14
Quartz sand	Proppant	3,13
Quaternary amine		8
Quaternary amine compounds		12
Quaternary ammonium compound		8,12

Table continued on next page

Table E1 continued from previous page

Chemical Name	Use	Ref.
Quaternary ammonium compounds, (oxydi-2,1-ethanediyl)bis[coco alkyldimethyl, dichlorides		7,14
Quaternary ammonium compounds, benzylbis(hydrogenated tallow alkyl)methyl, salts with bentonite	Fluid additives	5,6,13
Quaternary ammonium compounds, benzyl-C12-16-alkyldimethyl, chlorides		12
Quaternary ammonium compounds, bis(hydrogenated tallow alkyl)dimethyl, salts with bentonite		14
Quaternary ammonium compounds, bis(hydrogenated tallow alkyl)dimethyl, salts with hectorite	Viscosifier	13
Quaternary ammonium compounds, dicoco alkyldimethyl, chlorides		12
Quaternary ammonium compounds, trimethyltallow alkyl, chlorides		12
Quaternary ammonium salts		8,12,14
Quaternary compound		12
Quaternary salt		12,14
Radium (228)		4
Raffinates (petroleum)		5
Raffinates, petroleum, sorption process		12
Residual oils, petroleum, solvent-refined		5
Residues, petroleum, catalytic reformer fractionator		12,14
Resin		14
Rosin		12
Rutile		12
Saline	Brine carrier fluid, breaker	5,10,12,13,14
Salt		14
Salt of amine-carbonyl condensate		14
Salt of fatty acid/polyamine reaction product		14
Salt of phosphate ester		12
Salt of phosphono-methylated diamine		12
Salts of alkyl amines	Foaming agent	1,6,13
Sand		14
Saturated sucrose		7,12,14
Secondary alcohol		12
Selenium		7
Sepiolite		14
Silane, dichlorodimethyl-, reaction products with silica		14
Silica	Proppant	3,12,13,14
Silica gel, cryst.-free		14
Silica, amorphous		12
Silica, amorphous precipitated		12,14
Silica, microcrystalline		13
Silica, quartz sand		14
Silicic acid (H_4SiO_4), tetramethyl ester		12
Silicon dioxide (fused silica)		12,14

Table continued on next page

Table E1 continued from previous page

Chemical Name	Use	Ref.
Silicone emulsion		12
Silicone ester		14
Silver		7
Silwet L77		12
Soda ash		14
Sodium		4
Sodium 1-octanesulfonate		7,14
Sodium 2-mercaptobenzothiolate	Corrosion inhibitor	13
Sodium acetate		7,12,14
Sodium alpha-olefin Sulfonate		14
Sodium aluminum oxide		12
Sodium benzoate		7,14
Sodium bicarbonate		5,9,12,14
Sodium bisulfite, mixture of $NaHSO_3$ and $Na_2S_2O_5$		7,12,14
Sodium bromate	Breaker	12,13,14
Sodium bromide		7,9,12,14
Sodium carbonate	pH control	3,12,13,14
Sodium chlorate		12,14
Sodium chlorite	Breaker	7,10,12,13,14
Sodium chloroacetate		7,14
Sodium cocaminopropionate		12
Sodium decyl sulfate		12
Sodium diacetate		12
Sodium dichloroisocyanurate	Biocide	13
Sodium erythorbate		7,12,14
Sodium ethasulfate		12
Sodium formate		14
Sodium hydroxide	Gelling agent	1,9,12,13,14
Sodium hypochlorite		7,12,14
Sodium iodide		14
Sodium ligninsulfonate	Surfactant	13
Sodium metabisulfite		12
Sodium metaborate		7,12,14
Sodium metaborate tetrahydrate		12
Sodium metasilicate		12,14
Sodium nitrate	Fluid additives	13
Sodium nitrite	Corrosion inhibitor	12,13,14
Sodium octyl sulfate		12
Sodium oxide (Na_2O)		12
Sodium perborate		12
Sodium perborate tetrahydrate	Concentrate	7,10,12,13,14
Sodium persulfate		5,9,12,14
Sodium phosphate		12,14
Sodium polyacrylate		7,12,14
Sodium pyrophosphate		5,12,14
Sodium salicylate		12
Sodium silicate		12,14
Sodium sulfate		7,12,14

Table continued on next page

Table E1 continued from previous page

Chemical Name	Use	Ref.
Sodium sulfite		14
Sodium tetraborate decahydrate	Crosslinker	1,6,13
Sodium thiocyanate		12
Sodium thiosulfate		7,12,14
Sodium thiosulfate, pentahydrate		12
Sodium trichloroacetate		12
Sodium xylenesulfonate		9,12
Sodium zirconium lactate		12
Sodium α-olefin sulfonate		7
Solvent naphtha, petroleum, heavy aliph.		14
Solvent naphtha, petroleum, heavy arom.	Non-ionic surfactant	5,10,12,13,14
Solvent naphtha, petroleum, light arom.	Surfactant	12,13,14
Sorbitan, mono-(9Z)-9-octadecenoate		7,12,14
Stannous chloride dihydrate		12,14
Starch	Proppant	12,14
Starch blends	Fluid additives	6
Steam cracked distillate, cyclodiene dimer, dicyclopentadiene polymer		12
Steranes		4
Stoddard solvent		7,12,14
Stoddard solvent IIC		7,12,14
Strontium		7
Strontium (89&90)		13
Styrene	Proppant	13
Substituted alcohol		12
Substituted alkene		12
Substituted alkylamine		12
Sugar		14
Sulfamic acid		7,12,14
Sulfate		4,7,12,14
Sulfite		7
Sulfomethylated tannin		5
Sulfonate acids		12
Sulfonate surfactants		12
Sulfonic acid salts		12
Sulfonic acids, C14-16-alkane hydroxy and C14-16-alkene, sodium salts		7,12,14
Sulfonic acids, petroleum		12
Sulfur compound		12
Sulfuric acid		9,12,14
Surfactant blend		14
Surfactants		9,12
Symclosene		8
Synthetic organic polymer		12,14
Talc	Fluid additives	5,6,9,12,13,14
Tall oil, compound with diethanolamine		12
Tallow soap		12,14

Table continued on next page

Table E1 continued from previous page

Chemical Name	Use	Ref.
Tar bases, quinoline derivatives, benzyl chloride-quaternized		7,12,14
Tebuthiuron		8
Terpenes		12
Terpenes and terpenoids, sweet orange-oil		7,12,14
Terpineol, mixture of isomers		7,12,14
tert-Butyl hydroperoxide (70% solution in water)		12,14
tert-Butyl perbenzoate		12
Tetra-calcium-alumino-ferrite		12,14
Tetrachloroethylene		7
Tetradecyl dimethyl benzyl ammonium chloride		12
Tetraethylene glycol		12
Tetraethylenepentamine		12,14
Tetrakis(hydroxymethyl)phosphonium sulfate		7,9,12,14
Tetramethylammonium chloride		7,9,12,14
Thallium and compounds		7
Thiocyanic acid, ammonium salt		7,14
Thioglycolic acid	Iron Control	12,13,14
Thiourea	Acid corrosion inhibitor	1,6,12,13,14
Thiourea polymer		12,14
Thorium		2
Tin		1
Tin(II) chloride		12
Titanium	Crosslinker	4
Titanium complex		12,14
Titanium dioxide	Proppant	12,13,14
Titanium(4+) 2-[bis(2-hydroxyethyl)amino]ethanolate propan-2-olate (1:2:2)		12
Titanium, isopropoxy (triethanolaminate)		12
TOC		7
Toluene	Gelling agent	1,12,14
trans-Squalene		8
Tributyl phosphate	Defoamer	13
Tricalcium phosphate		12
Tricalcium silicate		12,14
Triethanolamine		5,12,14
Triethanolamine hydroxyacetate		7,14
Triethanolamine polyphosphate ester		12
Triethanolamine zirconium chelate		12
Triethyl citrate		12
Triethyl phosphate		12,14
Triethylene glycol		5,12,14
Triisopropanolamine		12,14
Trimethyl ammonium chloride		9,14
Trimethylamine quaternized polyepichlorohydrin		5,12,14
Trimethylbenzene	Fracturing fluid	12,13
Tri-n-butyl tetradecyl phosphonium chloride		7,12,14
Triphosphoric acid, pentasodium salt		12,14

Table continued on next page

Table E1 continued from previous page

Chemical Name	Use	Ref.
Tripropylene glycol monomethyl ether	Viscosifier	13
Tris(hydroxymethyl)amine		7
Trisodium citrate		7,14
Trisodium ethylenediaminetetraacetate		12,14
Trisodium ethylenediaminetriacetate		12
Trisodium phosphate		7,12,14
Trisodium phosphate dodecahydrate		12
Triterpanes		4
Triton X-100		7,12,14
Ulexite		12,14
Ulexite, calcined		14
Ultraprop		14
Undecane		7,14
Uranium-238		2
Urea		7,12,14
Vanadium		1
Vanadium compounds		14
Vermiculite	Lubricant	13
Versaprop		14
Vinylidene chloride/methylacrylate copolymer		14
Wall material		12
Walnut hulls		12,14
Water	Water gelling agent/ foaming agent	1,14
White mineral oil, petroleum		12,14
Xylenes	Gelling agent	1,12,14
Yttrium		1
Zinc	Lubricant	13
Zinc carbonate	Corrosion inhibitor	13
Zinc chloride		12
Zinc oxide		12
Zirconium		7
Zirconium complex	Crosslinker	5,10,12,14
Zirconium nitrate	Crosslinker	1,6
Zirconium oxide sulfate		12
Zirconium oxychloride	Crosslinker	12,13
Zirconium sodium hydroxy lactate complex (sodium zirconium lactate)		12
Zirconium sulfate	Crosslinker	1,6
Zirconium, acetate lactate oxo ammonium complexes		14
Zirconium,tetrakis[2-[bis(2-hydroxyethyl)amino-kN]ethanolato-kO]-	Crosslinker	10,12,14
α-[3.5-Dimethyl-1-(2-methylpropyl)hexyl]-w-hydroxy-poly(oxy-1,2-ethandiyl)		7,14

References

1. Sumi, L. (2005). *Our drinking water at risk. What EPA and the oil and gas industry don't want us to know about hydraulic fracturing.* Durango, CO: Oil and Gas Accountability Project/Earthworks. Retrieved January 21, 2011, from http://www.earthworksaction.org/pubs/DrinkingWaterAtRisk.pdf.

2. Sumi, L. (2008). *Shale gas: Focus on the Marcellus Shale.* Oil and Gas Accountability Project. Durango, CO.

3. Ground Water Protection Council & ALL Consulting. (2009). *Modern shale gas development in the US: A primer.* Washington, DC: US Department of Energy, Office of Fossil Energy and National Energy Technology Laboratory. Retrieved January 19, 2011, from http://www.netl.doe.gov/technologies/oil-gas/publications/EPreports/Shale_Gas_Primer_2009.pdf.

4. Veil, J. A., Puder, M. G., Elcock, D., & Redweik, R. J. (2004). *A white paper describing produced water from production of crude oil, natural gas, and coal bed methane.* Argonne National Laboratory Report for U.S. Department of Energy, National Energy Technology Laboratory.

5. Material Safety Data Sheets; EnCana Oil & Gas (USA), Inc.: Denver, CO. Provided by EnCana upon US EPA Region 8 request as part of the Pavillion, WY, ground water investigation.

6. US Environmental Protection Agency. (2004). *Evaluation of impacts to underground sources of drinking water by hydraulic fracturing of coalbed methane reservoirs.* No. EPA/816/R-04/003. Washington, DC: US Environmental Protection Agency, Office of Water.

7. New York State Department of Environmental Conservation. (2009, September). *Supplemental generic environmental impact statement on the oil, gas and solution mining regulatory program (draft).* Well permit issuance for horizontal drilling and high-volume hydraulic fracturing to develop the Marcellus Shale and other low-permeability gas reservoirs. Albany, NY: New York State Department of Environmental Conservation. Retrieved January 20, 2010, from ftp://ftp.dec.state.ny.us/dmn/download/OGdSGEISFull.pdf.

8. US Environmental Protection Agency.(2010). *Region 8 analytical lab analysis.*

9. Bureau of Oil and Gas Management. (2010). *Chemicals used in the hydraulic fracturing process in Pennsylvania.* Pennsylvania Department of Environmental Protection. Retrieved September 12, 2011, from http://assets.bizjournals.com/cms_media/pittsburgh/datacenter/DEP_Frac_Chemical_List_6-30-10.pdf.

10. Material Safety Data Sheets; Halliburton Energy Services, Inc.: Duncan, OK. Provided by Halliburton Energy Services during an on-site visit by EPA on May 10, 2010.

11. Alpha Environmental Consultants, Inc., Alpha Geoscience, NTS Consultants, Inc. (2009). *Issues related to developing the Marcellus Shale and other low-permeability gas reservoirs.* Report for the New York State Energy Research and Development Authority. NYSERDA Contract No. 11169, NYSERDA Contract No. 10666, and NYSERDA Contract No. 11170. Albany, NY.

12. US House of Representatives Committee on Energy and Commerce Minority Staff (2011). *Chemicals used in hydraulic fracturing.*

13. US Environmental Protection Agency. (2010). *Expanded site investigation analytical report: Pavillion Area groundwater investigation.* Contract No. EP-W-05-050. Retrieved September 7, 2011, from http://www.epa.gov/region8/superfund/wy/pavillion/PavillionAnalyticalResultsReport.pdf.

14. Submitted non-Confidential Business Information by Halliburton, Patterson and Superior. Available on the Federal Docket, EPA-HQ-ORD-2010-0674.

TABLE E2. CHEMICALS IDENTIFIED IN FLOWBACK/PRODUCED WATER

Chemical	Ref.	Chemical	Ref.
1,1,1-Trifluorotoluene	1	Atrazine	3
1,2-Bromo-2-nitropropane-1,3-diol (2-bromo-2-nitro-1,3-propanediol or bronopol)	3	Barium	2
		Bentazon	3
1,-3-Dimethyladamantane	3	Benzene	2
1,4-Dichlorobutane	1	Benzo(a)pyrene	2
1,6-Hexanediamine	3	Benzyldimethyl-(2-prop-2-enoyloxyethyl)ammonium chloride	3
1-Methoxy-2-propanol	3		
2-(2-Methoxyethoxy)ethanol	3		
2-(Thiocyanomethylthio)benzothiazole	3	Benzylsuccinic acid	3
		Beryllium	4
2,2,2-Nitrilotriethanol	3	Bicarbonate	1
2,2-Dibromo-3-nitrilopropionamide	3	Bis(2-ethylhexyl)phthalate	1
		Bis(2-ethylhexyl)phthalate	4
2,2-Dibromoacetonitrile	3	Bisphenol a	3
2,2-Dibromopropanediamide	3	Boric acid	3
2,4,6-Tribromophenol	1	Boric oxide	3
2,4-Dimethylphenol	2	Boron	1,2
2,5-Dibromotoluene	1	Bromide	1
2-Butanone	2	Bromoform	1
2-Butoxyacetic acid	3	Butanol	3
2-Butoxyethanol	3	Cadmium	2
2-Butoxyethanol phosphate	3	Calcium	2
2-Ethyl-3-propylacrolein	3	Carbonate alkalinity	1
2-Ethylhexanol	3	Cellulose	3
2-Fluorobiphenyl	1	Chloride	2
2-Fluorophenol	1	Chlorobenzene	2
3,5-Dimethyl-1,3,5-thiadiazinane-2-thione	3	Chlorodibromomethane	1
		Chloromethane	4
4-Nitroquinoline-1-oxide	1	Chrome acetate	3
4-Terphenyl-d14	1	Chromium	4
5-Chloro-2-methyl-4-isothiazolin-3-one	3	Chromium hexavalent	
		Citric acid	3
6-Methylquinoline	3	Cobalt	1
Acetic acid	3	Copper	2
Acetic anhydride	3	Cyanide	1
Acrolein	3	Cyanide	4
Acrylamide (2-propenamide)	3	Decyldimethyl amine	3
Adamantane	3	Decyldimethyl amine oxide	3
Adipic acid	3	Diammonium phosphate	3
Aluminum	2	Dichlorobromomethane	1
Ammonia	4	Didecyl dimethyl ammonium chloride	3
Ammonium nitrate	3		
Ammonium persulfate	3	Diethylene glycol	3
Anthracene	2	Diethylene glycol monobutyl ether	3
Antimony	1		
Arsenic	2	Dimethyl formamide	3

Table continued on next page

Table E2 continued from previous page

Chemical	Ref.	Chemical	Ref.
Dimethyldiallylammonium chloride	3	Methylene phosphonic acid (diethylenetriaminepenta[methyl enephosphonic] acid)	3
Di-n-butylphthalate	2	Modified polysaccharide or pregelatinized cornstarch or starch	3
Dipropylene glycol monomethyl ether	3		
Dodecylbenzene sulfonic acid	3	Molybdenum	1
Eo-C7-9-iso-,C8 rich-alcohols	3	Monoethanolamine	3
Eo-C9-11-iso, C10-rich alcohols	3	Monopentaerythritol	3
Ethoxylated 4-nonylphenol	3	m-Terphenyl	3
Ethoxylated nonylphenol	3	Muconic acid	3
Ethoxylated nonylphenol (branched)	3	N,N,N-trimethyl-2[1-oxo-2-propenyl]oxy ethanaminium chloride	3
Ethoxylated octylphenol	3	n-Alkanes, C10-C18	2
Ethyl octynol	3	n-Alkanes, C18-C70	2
Ethylbenzene	2	n-Alkanes, C1-C2	2
Ethylbenzene	3	n-Alkanes, C2-C3	2
Ethylcellulose	3	n-Alkanes, C3-C4	2
Ethylene glycol	3	n-Alkanes, C4-C5	2
Ethylene glycol monobutyl ether	3	n-Alkanes, C5-C8	2
Ethylene oxide	3	Naphthalene	2
Ferrous sulfate heptahydrate	3	Nickel	2
Fluoride	1	Nitrazepam	3
Formamide	3	Nitrobenzene	3
Formic acid	3	Nitrobenzene-d5	1
Fumaric acid	3	n-Methyldiethanolamine	3
Glutaraldehyde	3	Oil and grease	2
Glycerol	3	o-Terphenyl	1
Hydroxyethylcellulose	3	o-Terphenyl	3
Hydroxypropylcellulose	3	Oxiranemethanaminium, N,N,N-trimethyl-, chloride, homopolymer	3
Iron	2		
Isobutyl alcohol (2-methyl-1-propanol)	3	p-Chloro-m-cresol	2
Isopropanol (propan-2-ol)	3	Petroleum hydrocarbons	1
Lead	2	Phenol	2
Limonene	3	Phosphonium, tetrakis(hydroxymethly)-sulfate	3
Lithium	1		
Magnesium	2	Phosphorus	1
Manganese	2	Polyacrylamide	3
Mercaptoacidic acid	3	Polyacrylate	3
Mercury	4	Polyethylene glycol	3
Methanamine,N,N-dimethyl-,N-oxide	3	Polyhexamethylene adipamide	3
		Polypropylene glycol	3
Methanol	3	Polyvinyl alcohol [alcotex 17f-h]	3
Methyl bromide	1	Potassium	1
Methyl chloride	1	Propane-1,2-diol	3
Methyl-4-isothiazolin	3		
Methylene bis(thiocyanate)	3		

Table continued on next page

Table E2 continued from previous page

Chemical	Ref.
Propargyl alcohol	3
Pryidinium, 1-(phenylmethyl)-, ethyl methyl derivatives, chlorides	3
p-Terphenyl	3
Quaternary amine	3
Quaternary ammonium compound	3
Quaternary ammonium salts	3
Radium (226)	2
Radium (228)	2
Selenium	1
Silver	1
Sodium	2
Sodium carboxymethylcellulose	3
Sodium dichloro-s-triazinetrione	3
Sodium mercaptobenzothiazole	3
Squalene	3
Steranes	2
Strontium	1
Sucrose	3
Sulfate	1,2
Sulfide	1
Sulfite	1
Tebuthiuron	3
Terpineol	3
Tetrachloroethene	4
Tetramethyl ammonium chloride	3
Tetrasodium ethylenediaminetetraacetate	3
Thallium	1
Thiourea	3
Titanium	2
Toluene	2
Total organic carbon	1
Tributyl phosphate	3
Trichloroisocyanuric acid	3
Trimethylbenzene	3
Tripropylene glycol methyl ether	3
Trisodium nitrilotriacetate	3
Triterpanes	2
Urea	3
Xylene (total)	2
Zinc	2
Zirconium	1

References

1. New York State Department of Environmental Conservation. (2011, September). *Supplemental generic environmental impact statement on the oil, gas and solution mining regulatory program (draft). Well permit issuance for horizontal drilling and high-volume hydraulic fracturing to develop the Marcellus Shale and other low-permeability gas reservoirs.* Albany, NY: New York State Department of Environmental Conservation. Retrieved January 20, 2010, from ftp://ftp.dec.state.ny.us/dmn/download/OGdSGEISFull.pdf.

2. Veil, J. A., Puder, M. G., Elcock, D., & Redweik, R. J. (2004). *A white paper describing produced water from production of crude oil, natural gas, and coal bed methane.* Prepared for the US Department of Energy, National Energy Technology Laboratory. Argonne, IL: Argonne National Laboratory. Retrieved January 20, 2011, from http://www.evs.anl.gov/pub/doc/ProducedWatersWP0401.pdf.

3. URS Operating Services, Inc. (2010, August 20). *Expanded site investigation—Analytical results report. Pavillion area groundwater investigation.* Prepared for US Environmental Protection Agency. Denver, CO: URS Operating Services, Inc. Retrieved January 27, 2011, from http://www.epa.gov/region8/superfund/wy/pavillion/PavillionAnalyticalResultsReport.pdf.

4. Alpha Environmental Consultants, Inc., Alpha Geoscience, & NTS Consultants, Inc. (2009). *Issues related to developing the Marcellus Shale and other low-permeability gas reservoirs.* Albany, NY: New York State Energy Research and Development Authority.

TABLE E3. NATURALLY OCCURRING SUBSTANCES MOBILIZED BY FRACTURING ACTIVITIES

Chemical	Common Valence States	Ref.
Aluminum	III	1
Antimony	V,III,-III	1
Arsenic	V, III, 0, -III	1
Barium	II	1
Beryllium	II	1
Boron	III	1
Cadmium	II	1
Calcium	II	1
Chromium	VI, III	1
Cobalt	III, II	1
Copper	II, I	1
Hydrogen sulfide	N/A	2
Iron	III, II	1
Lead	IV, II	1
Magnesium	II	1
Molybdenum	VI, III	1
Nickel	II	1
Radium (226)	II	2
Radium (228)	II	2
Selenium	VI, IV, II, 0, -II	1
Silver	I	1
Sodium	I	1
Thallium	III, I	1
Thorium	IV	2
Tin	IV, II, -IV	1
Titanium	IV	1
Uranium	VI, IV	2
Vanadium	V	1
Yttrium	III	1
Zinc	II	1

References

1. Sumi, L. (2005). *Our drinking water at risk: What EPA and the oil and gas industry don't want us to know about hydraulic fracturing*. Durango, CO: Oil and Gas Accountability Project/Earthworks. Retrieved January 21, 2011, from http://www.earthworksaction.org/pubs/DrinkingWaterAtRisk.pdf.

2. Sumi, L. (2008). *Shale gas: Focus on the Marcellus Shale*. Durango, CO: Oil and Gas Accountability Project/Earthworks. Retrieved January 21, 2011, from http://www.earthworksaction.org/pubs/OGAPMarcellusShaleReport-6-12-08.pdf.

APPENDIX F: STAKEHOLDER-NOMINATED CASE STUDIES

This appendix lists the stakeholder-nominated case studies. Potential retrospective case study sites can be found in Table F1, while potential prospective case study sites are listed in Table F2.

TABLE F1. POTENTIAL RETROSPECTIVE CASE STUDY SITES

Formation	Location	Key Areas to Be Addressed	Key Activities	Potential Outcomes	Partners
Bakken Shale	Killdeer and Dunn Co., ND	Production well failure during hydraulic fracturing; suspected drinking water aquifer contamination; surface waters nearby; soil contamination; more than 2,000 barrels of oil and fracturing fluids leaked from the well	Monitoring wells to evaluate extent of contamination of aquifer; soil and surface water monitoring	Determine extent of contamination of drinking water resources; identify sources of well failure	NDDMR-Industrial Commission, EPA Region 8, Berthold Indian Reservation
Barnett Shale	Alvord, TX	Benzene in water well			RRCTX, landowners, USGS, EPA Region 6
Barnett Shale	Azle, TX	Skin rash complaints from contaminated water			RRCTX, landowners, USGS, EPA Region 6
Barnett Shale	Decatur, TX	Skin rash complaints from drilling mud applications to land			RRCTX, landowners, USGS, EPA Region 6

Table continued on next page

Table F1 continued from previous page

Formation	Location	Key Areas to Be Addressed	Key Activities	Potential Outcomes	Partners
Barnett Shale	Wise/Denton Cos. (including Dish), TX	Potential drinking water well contamination; surface spills; waste pond overflow; documented air contamination	Monitor other wells in area and install monitoring wells to evaluate source(s)	Determine sources of contamination of private well	RRCTX, TCEQ, landowners, City of Dish, USGS, EPA Region 6, DFW Regional Concerned Citizens Group, North Central Community Alliance, Sierra Club
Barnett Shale	South Parker Co. and Weatherford, TX	Hydrocarbon contamination in multiple drinking water wells; may be from faults/fractures from production well beneath properties	Monitor other wells in area; install monitoring wells to evaluate source(s)	Determine source of methane and other contaminants in private water well; information on role of fracture/fault pathway from hydraulic fracturing zone	RRCTX, landowners, USGS, EPA Region 6
Barnett Shale	Tarrant Co., TX	Drinking water well contamination; report of leaking pit	Monitoring well	Determine if pit leak impacted underlying ground water	RRCTX, landowners, USGS, EPA Region 6
Barnett Shale	Wise Co. and Decatur, TX	Spills; runoff; suspect drinking water well contamination; air quality impacts	Sample wells, soils	Determine sources of contamination of private well	RRCTX, landowners, USGS, EPA Region 6, Earthworks Oil & Gas Accountability Project
Clinton Sandstone	Bainbridge, OH	Methane buildup leading to home explosion			OHDNR, EPA Region 5
Fayetteville Shale	Arkana Basin, AR	General water quality concerns			AROGC, ARDEQ, EPA Region 6
Fayetteville Shale	Conway Co., AR	Gray, smelly water			AROGC, ARDEQ, EPA Region 6

Table continued on next page

Table F1 continued from previous page

Formation	Location	Key Areas to Be Addressed	Key Activities	Potential Outcomes	Partners
Fayetteville Shale	Van Buren or Logan Cos., AR	Stray gas (methane) in wells; other water quality impairments			AROGC, ARDEQ, EPA Region 6
Haynesville Shale	Caddo Parish, LA	Drinking water impacts (methane in water)	Monitoring wells to evaluate source(s)	Evaluate extent of water well contamination and if source is from hydraulic fracturing operations	LGS, USGS, EPA Region 6
Haynesville Shale	DeSoto Parish, LA	Drinking water reductions	Monitoring wells to evaluate water availability; evaluate existing data	Determine source of drinking water reductions	LGS, USGS, EPA Region 6
Haynesville Shale	Harrison Co., TX	Stray gas in water wells			RRCTX, landowners, USGS, EPA Region 6
Marcellus Shale	Bradford Co., PA	Drinking water well contamination; surface spill of hydraulic fracturing fluids	Soil, ground water, and surface water sampling	Determine source of methane in private wells	PADEP, landowners, EPA Region 3, Damascus Citizens Group, Friends of the Upper Delaware
Marcellus Shale	Clearfield Co., PA	Well blowout			PADEP, EPA Region 3
Marcellus Shale	Dimock, Susquehanna Co., PA	Contamination in multiple drinking water wells; surface water quality impairment from spills	Soil, ground water, and surface water sampling	Determine source of methane in private wells	PADEP, EPA Region 3, landowners, Damascus Citizens Group, Friends of the Upper Delaware
Marcellus Shale	Gibbs Hill, PA	On-site spills; impacts to drinking water; changes in water quality	Evaluate existing data; determine need for additional data	Evaluate extent of large surface spill's impact on soils, surface water, and ground water	PADEP, landowner, EPA Region 3

Table continued on next page

154

Table F1 continued from previous page

Formation	Location	Key Areas to Be Addressed	Key Activities	Potential Outcomes	Partners
Marcellus Shale	Hamlin Township and McKean Co., PA	Drinking water contamination from methane; changes in water quality	Soil, ground water, and surface water sampling	Determine source of methane in community and private wells	PADEP, EPA Region 3, Schreiner Oil & Gas
Marcellus Shale	Hickory, PA	On-site spill; impacts to drinking water; changes in water quality; methane in wells; contaminants in drinking water (acrylonitrile, VOCs)			PADEP, landowner, EPA Region 3
Marcellus Shale	Hopewell Township, PA	Surface spill of hydraulic fracturing fluids; waste pit overflow	Sample pit and underlying soils; sample nearby soil, ground water, and surface water	Evaluate extent of large surface spill's impact on soils, surface water, and ground water	PADEP, landowners, EPA Region 3
Marcellus Shale	Indian Creek Watershed, WV	Concerns related to wells in karst formation			WVOGCC, EPA Region 3
Marcellus Shale	Lycoming Co., PA	Surface spill of hydraulic fracturing fluids	PADEP sampled soils, nearby surface water, and two nearby private wells; evaluate need for additional data collection to determine source of impact	Evaluate extent of large surface spill's impact on soils, surface water, and ground water	
Marcellus Shale	Monongahela River Basin, PA	Surface water impairment (high TDS, water availability)	Data exists on water quality over time for Monongahela River during ramp up of hydraulic fracturing activity; review existing data	Assess intensity of hydraulic fracturing activity	
Marcellus Shale	Susquehanna River Basin, PA and NY	Water availability; water quality	Assess water use and water quality over time; review existing data	Determine if water withdrawals for hydraulic fracturing are related to changes in water quality and availability	
Marcellus Shale	Tioga Co., NY	General water quality concerns			
Marcellus Shale	Upshur Co., WV	General water quality concerns			WVOGCC, EPA Region 3

Table continued on next page

Table F1 continued from previous page

Formation	Location	Key Areas to Be Addressed	Key Activities	Potential Outcomes	Partners
Marcellus Shale	Wetzel Co., WV, and Washington/ Green Cos., PA	Stray gas; spills; changes in water quality; several landowners concerned about methane in wells	Soil, ground water, and surface water sampling	Determine extent of impact from spill of hydraulic fracturing fluids associated with well blowout and other potential impacts to drinking water resources	WVDEP, WVOGCC, PADEP, EPA Region 3, landowners, Damascus Citizens Group
Piceance Basin	Battlement Mesa, CO	Water quality and quantity concerns			COGCC, landowners, EPA Region 8
Piceance Basin (tight gas sand)	Garfield Co., CO (Mamm Creek area)	Drinking water well contamination; changes in water quality; water levels	Soil, ground water, and surface water sampling; review existing data	Evaluate source of methane and degradation in water quality basin-wide	COGCC, landowners, EPA Region 8, Colorado League of Women Voters
Piceance Basin	Rifle, CO	Water quality and quantity concerns			COGCC, landowners, EPA Region 8
Piceance Basin	Silt, CO	Water quality and quantity concerns			COGCC, landowners, EPA Region 8
Powder River Basin (CBM)	Clark, WY	Drinking water well contamination	Monitoring wells to evaluate source(s)	Evaluate extent of water well contamination and if source is from hydraulic fracturing operations	WOOGC, EPA Region 8, landowners
San Juan Basin (shallow CBM and tight sand)	LaPlata Co., CO	Drinking water well contamination, primarily with methane (area along the edge of the basin has large methane seepage)	Large amounts of data have been collected through various studies of methane seepage; gas wells at the margin of the basin can be very shallow	Evaluate extent of water well contamination and determine if hydraulic fracturing operations are the source	COGCC, EPA Region 8, BLM, San Juan Citizens Alliance

Table continued on next page

Table F1 continued from previous page

Formation	Location	Key Areas to Be Addressed	Key Activities	Potential Outcomes	Partners
Raton Basin (CBM)	Huerfano Co., CO	Drinking water well contamination; methane in well water; well house explosion	Monitoring wells to evaluate source of methane and degradation in water quality	Evaluate extent of water well contamination and determine if hydraulic fracturing operations are the source	COGCC, EPA Region 8
Raton Basin (CBM)	Las Animas Co., CO	Concerns about methane in water wells			COGCC, landowners, EPA Region 8
Raton Basin (CBM)	North Fork Ranch, Las Animas Co., CO	Drinking water well contamination; changes in water quality and quantity	Monitoring wells to evaluate source of methane and degradation in water quality	Evaluate extent of water well contamination and determine if hydraulic fracturing operations are the source	COGCC, landowners, EPA Region 8
Tight gas sand	Garfield Co., CO	Drinking water and surface water contamination; documented benzene contamination	Monitoring to assess source of contamination	Determine if contamination is from hydraulic fracturing operations in area	COGCC, EPA Region 8, Battlement Mesa Citizens Group
Tight gas sand	Pavillion, WY	Drinking water well contamination	Monitoring wells to evaluate source(s) (ongoing studies by ORD and EPA Region 8)	Determine if contamination is from hydraulic fracturing operations in area	WOGCC, EPA Region 8, landowners
Tight gas sand	Sublette Co., WY (Pinedale Anticline)	Drinking water well contamination (benzene)	Monitoring wells to evaluate source(s)	Evaluate extent of water well contamination and determine if hydraulic fracturing operations are the source	WOGCC, EPA Region 8, Earthworks

Within the scope of this study, prospective case studies will focus on key areas such as the full lifecycle and environmental monitoring. To address these issues, key research activities will include water and soil monitoring before, during, and after hydraulic fracturing activities.

TABLE F2. PROSPECTIVE CASE STUDIES

Formation	Location	Potential Outcomes	Partners
Bakken Shale	Berthold Indian Reservation, ND	Baseline water quality data, comprehensive monitoring and modeling of water resources during all stages of the hydraulic fracturing process	NDDMR-Industrial Commission, University of North Dakota, EPA Region 8, Berthold Indian Reservation
Barnett Shale	Flower Mound/ Bartonville, TX	Baseline water quality data, comprehensive monitoring and modeling of water resources during all stages of the hydraulic fracturing process	NDDMR-Industrial Commission, EPA Region 8, Mayor of Flower Mound
Marcellus Shale	Otsego Co., NY	Baseline water quality data, comprehensive monitoring and modeling of water resources during all stages of the hydraulic fracturing process	NYSDEC; Gastem, USA; others TBD
Marcellus Shale	TBD, PA	Baseline water quality data, comprehensive monitoring and modeling of water resources during all stages of the hydraulic fracturing process in a region of the country experiencing intensive hydraulic fracturing activity	Chesapeake Energy, PADEP, others TBD
Marcellus Shale	Wyoming Co, PA	Baseline water quality data, comprehensive monitoring and modeling of water resources during all stages of the hydraulic fracturing process	DOE, PADEP, University of Pittsburgh, Range Resources, USGS, landowners, EPA Region 3
Niobrara Shale	Laramie Co., WY	Baseline water quality data, comprehensive monitoring and modeling of water resources during all stages of the hydraulic fracturing process, potential epidemiology study by Wyoming Health Department	WOGCC, Wyoming Health Department, landowners, USGS, EPA Region 8
Woodford Shale or Barnett Shale	OK or TX	Baseline water quality data, comprehensive monitoring and modeling of water resources during all stages of the hydraulic fracturing process	OKCC, landowners, USGS, EPA Region 6

Appendix F Acronym List

ARDEQ	Arkansas Department of Environmental Quality
AROGC	Arkansas Oil and Gas Commission
BLM	Bureau of Land Management
CBM	coalbed methane
Co.	county
COGCC	Colorado Oil and Gas Conservation Commission
DFW	Dallas-Fort Worth
DOE	US Department of Energy
EPA	US Environmental Protection Agency
LGS	Louisiana Geological Survey
NDDMR	North Dakota Department of Mineral Resources
NYSDEC	New York Department of Environmental Conservation
OHDNR	Ohio Department of Natural Resources
OKCC	Oklahoma Corporation Commission
PADEP	Pennsylvania Department of Environmental Protection
RRCTX	Railroad Commission of Texas
TBD	to be determined
TCEQ	Texas Commission on Environmental Quality
USACE	US Army Corps of Engineers
USGS	US Geological Survey
VOC	volatile organic compound
WOGCC	Wyoming Oil and Gas Conservation Commission
WVDEP	West Virginia Department of Environmental Protection
WVOGCC	West Virginia Oil and Gas Conservation Commission

APPENDIX G: ASSESSING MECHANICAL INTEGRITY

In relation to hydrocarbon production, it is useful to distinguish between the internal and external mechanical integrity of wells. Internal mechanical integrity is concerned with the containment of fluids within the confines of the well. External mechanical integrity is related to the potential movement of fluids along the wellbore outside the well casing.

A well's mechanical integrity can be determined most accurately through a combination of data and tests that individually provide information, which can then be compiled and evaluated. This appendix provides a brief overview of the tools used to assess mechanical well integrity.

CEMENT BOND TOOLS

The effectiveness of the cementing process is determined using cement bond tools and/or cement evaluation tools. Cement bond tools are acoustic devices that produce data (cement bond logs) used to evaluate the presence of cement behind the casing. Cement bond logs generally include a gamma-ray curve and casing collar locator; transit time, which measures the time it takes for a specific sound wave to travel from the transmitter to the receiver; amplitude curve, which measures the strength of the first compressional cycle of the returning sound wave; and a graphic representation of the waveform, which displays the manner in which the received sound wave varies with time. This latter presentation, the variable density log, reflects the material through which the signal is transmitted. To obtain meaningful data, the tool must properly calibrated and be centralized in the casing to obtain data that is meaningful for proper evaluation of the cement behind the casing.

Other tools available for evaluating cement bonding use ultrasonic transducers arranged in a spiral around the tool or in a single rotating hub to survey the circumference of the casing. The transducers emit ultrasonic pulses and measure the received ultrasonic waveforms reflected from the internal and external casing interfaces. The resulting logs produce circumferential visualizations of the cement bonds with the pipe and borehole wall. Cement bonding to the casing can be measured quantitatively, while bonding to the formation can only be measured qualitatively. Even though cement bond/evaluation tools do not directly measure hydraulic seal, the measured bonding qualities do provide inferences of sealing.

The cement sheath can fail during well construction if the cement fails to adequately encase the well casing or becomes contaminated with drilling fluid or formation material. After a well has been constructed, cement sheath failure is most often related to temperature- and pressure-induced stresses resulting from operation of the well (Ravi et al., 2002). Such stresses can result in the formation of a microannulus, which can provide a pathway for the migration of fluids from high-pressure zones.

TEMPERATURE LOGGING

Temperature logging can be used to determine changes that have taken place in and adjacent to injection/production wells. The temperature log is a continuous recording of temperature versus depth.

Under certain conditions the tool can be used to conduct a flow survey, locating points of inflow or outflow in a well; locate the top of the cement in wells during the cement curing process (using the heat of hydration of the cement); and detect the flow of fluid and gas behind the casing. The temperature logging tool is the oldest of the production tools and one of the most versatile, but a highly qualified expert must use it and interpret its results.

NOISE LOGGING

The noise logging tool may have application in certain conditions to detect fluid movement within channels in cement in the casing/borehole annulus. It came into widespread application as a way to detect the movement of gas through liquid. For other flows, for example water through a channel, the tool relies on the turbulence created as the water flows through a constriction that creates turbulent flow. Two advantages of using the tool are its sensitivity and lateral depth of investigation. It can detect sound through multiple casings, and an expert in the interpretation of noise logs can distinguish flow behind pipe from flow inside pipe.

PRESSURE TESTING

A number of pressure tests are available to assist in determining the internal mechanical integrity of production wells. For example, while the well is being constructed, before the cement plug is drilled out for each casing, the casing should be pressure-tested to find any leaks. The principle of such a "standard pressure test" is that pressure applied to a fixed-volume enclosed vessel, closed at the bottom and the top, should remain constant if there are no leaks. The same concept applies to the "standard annulus pressure test," which is used when tubing and packers are a part of the well completion.

The "Ada" pressure test is used in some cases where the well is constructed with tubing without a packer, in wells with only casing and open perforations, and in dual injection/production wells.

The tools discussed above are summarized below in Table G1.

TABLE G1. COMPARISON OF TOOLS USED TO EVALUATE WELL INTEGRITY

Type of Tool	Description and Application	Types of Data
Acoustic cement bond tools	Acoustic devices to evaluate the presence of cement behind the casing	• Gamma-ray curve • Casing collar locator: depth control • Transit time: time it takes for a specific sound wave to travel from the transmitter to the receiver • Amplitude curve: strength of the first compressional cycle of the returning sound wave • Waveform: variation of received sound wave over time • Variable density log: reflects the material through which the signal is transmitted
Ultrasonic transducers	Transmit ultrasonic pulses and measure the received ultrasonic waveforms reflected from the internal and external casing interfaces to survey well casing	• Circumferential visualizations of the cement bonds with the pipe and borehole wall • Quantitative measures of cement bonding to the casing • Qualitative measure of bonding to the formation • Inferred sealing integrity
Temperature logging	Continuous recording of temperature versus depth to detect changes in and adjacent to injection/production wells	• Flow survey • Points of inflow or outflow in a well • Top of cement in wells during the cement curing process (using the heat of hydration of the cement) • Flow of fluid and gas behind casing
Noise logging tool	Recording of sound patterns that can be correlated to fluid movement; sound can be detected through multiple casings	• Fluid movement within channels in cement in the casing/borehole annulus
Pressure tests	Check for leaks in casing	• Changes in pressure within a fixed-volume enclosed vessel, implying that leaks are present

References

Ravi, K., Bosma, M., & Gastebled, O. (2002, April 30-May 2). *Safe and economic gas wells through cement design for life of the well.* No. SPE 75700. Presented at the Society of Petroleum Engineers Gas Technology Symposium, Calgary, Alberta, Canada.

APPENDIX H: FIELD SAMPLING AND ANALYTICAL METHODS

Field samples and monitoring data associated with hydraulic fracturing activities are collected for a variety of reasons, including to:

- Develop baseline data prior to fracturing.
- Monitor any changes in drinking water resources during and after hydraulic fracturing.
- Identify and quantify environmental contamination that may be associated with hydraulic fracturing.
- Evaluate well mechanical integrity.
- Evaluate the performance of treatment systems.

Field sampling is important for both the prospective and retrospective case studies discussed in Chapter 9. In retrospective case studies, EPA will take field samples to determine the cause of reported drinking water contamination. In prospective case studies, field sampling and monitoring provides for the identification of baseline conditions of the site prior to drilling and fracturing. Additionally, data will be collected during each step in the oil or natural gas drilling operation, including hydraulic fracturing of the formation and oil or gas production, which will allow EPA to monitor changes in drinking water resources as a result of hydraulic fracturing.

The case study site investigations will use monitoring wells and other available monitoring points to identify (and determine the quantity of) chemical compounds relevant to hydraulic fracturing activities in the subsurface environment. These compounds may include the chemical additives found in hydraulic fracturing fluid and their reaction/degradation products, as well as naturally occurring materials (e.g., formation fluid, gases, trace elements, radionuclides, and organic material) released during fracturing events.

This appendix first describes types of samples (and analytes associated with those samples) that may be collected throughout the oil and natural gas production process and the development and refinement of laboratory-based analytical methods. It then discusses the potential challenges associated with analyzing the collected field samples. The appendix ends with a summary of the data analysis process as well as a discussion of the evaluation of potential indicators associated with hydraulic fracturing activities.

FIELD SAMPLING: SAMPLE TYPES AND ANALYTICAL FOCUS

Table H1 lists monitoring and measurement parameters for both retrospective and prospective case studies. Note that samples taken in retrospective case studies will be collected after hydraulic fracturing has occurred and will focus on collecting evidence of contamination of drinking water resources. Samples taken for prospective case studies, however, will be taken during all phases of oil and gas production and will focus on improving EPA's understanding of hydraulic fracturing activities.

TABLE H1. MONITORING AND MEASUREMENT PARAMETERS AT CASE STUDY SITES

Sample Type	Case Study Site	Parameters
Surface and ground water (e.g., existing wells, new wells) Soil/sediments, soil gas	Prospective and retrospective (collect as much historical data as available)	• General water quality (e.g., pH, redox, dissolved oxygen) and water chemistry parameters (e.g., cations and anions) • Dissolved gases (e.g., methane) • Stable isotopes (e.g., Sr, Ra, C, H) • Metals • Radionuclides • Volatile and semi-volatile organic compounds, polycyclic aromatic hydrocarbons • Soil gas sampling in vicinity of proposed/actual hydraulic fracturing well location (e.g., Ar, He, H_2, O_2, N_2, CO_2, CH_4, C_2H_6, C_2H_4, C_3H_6, C_3H_8, iC_4H_{10}, nC_4H_{10}, iC_5H_{12})

Flowback and produced water	Prospective	• General water quality (e.g., pH, redox, dissolved oxygen, total dissolved solids) and water chemistry parameters (e.g., cations and anions)
		• Metals
		• Radionuclides
		• Volatile and semi-volatile organic compounds, polycyclic aromatic hydrocarbons
		• Sample fracturing fluids (time series sampling)
		   o Chemical concentrations
		   o Volumes injected
		   o Volumes recovered

Drill cuttings, core samples	Prospective	• Metals • Radionuclides • Mineralogic analyses

Table H1 indicates that field sampling will focus primarily on water and soil samples, which will be analyzed for naturally occurring materials and chemical additives used in hydraulic fracturing fluid, including their reaction products and/or degradates. Drill cuttings and core samples will be used in laboratory experiments to analyze the chemical composition of the formation and to explore chemical reactions between hydraulic fracturing fluid additives and the hydrocarbon-containing formation.

Data collected during the case studies are not restricted to the collection of field samples. Other data include results from mechanical integrity tests and surface geophysical testing. Mechanical well integrity can be assessed using a variety of tools, including acoustic cement bond tools, ultrasonic transducers, temperature and noise logging tools, and pressure tests. Geophysical testing can assess geologic and hydrogeologic conditions, detect and map underground structures, and evaluate soil and rock properties.

FIELD SAMPLING CONSIDERATIONS

Samples collected from drinking water taps or treatment systems will reflect the temperature, pressure, and redox conditions associated with the sampling site and may not reflect the true conditions in the subsurface, particularly in dissolved gas concentrations. In cases where dissolved gases are to be analyzed, special sampling precautions are needed. Because the depths of hydraulic fracturing wells can exceed 1,000 feet, ground water samples will be collected from settings where the temperature and

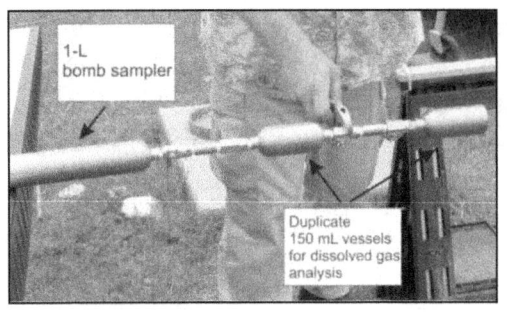

FIGURE H1. BOMB SAMPLER

pressure are significantly higher than at the surface. When liquid samples are brought to the surface, decreasing pressure can lead to off-gassing of dissolved gases (such as methane) and to changes in redox potential and pH that can lead to changes in the speciation and solubility of minerals and metals. Therefore, the sampling of water from these depths will require specialized sampling equipment that maintains the pressure of the formation until the sample is analyzed. One possible approach for this type of sampling is to employ a bomb sampler (shown in Figure G1) with a double-valve configuration that activates a series of stainless steel sampling vessels to collect pressurized ground water in one sampling pass.

USE OF PRESSURE TRANSDUCERS

Pressure transducers are a commonly used tool to measure water pressure changes correlated with changes in water levels within wells. The transducers are coupled with data loggers to electronically record the water level and time the measurement was obtained. They are generally used as an alternative to the frequent manual measurement of water levels. The devices used in this study consist of a small, self-contained pressure sensor, temperature sensor, battery, and non-volatile memory. The measurement frequency is programmable. Such data are often used to help predict groundwater flow directions and to evaluate possible relationships between hydraulic stresses (e.g., pumping, injection, natural recharge, etc.) and changes in water levels in wells, if sufficient data regarding the timing of the hydraulic stresses are available. These data may aid in evaluations of hydrostratigraphy and hydraulic communication within the aquifer.

DEVELOPMENT AND REFINEMENT OF LABORATORY-BASED ANALYTICAL METHODS

The ability to characterize chemical compounds related to hydraulic fracturing activities depends on the ability to detect and quantify individual constituents using appropriate analytical methods. As discussed in Chapter 6, EPA will identify the chemical additives used in hydraulic fracturing fluids as well as those found in flowback and produced water, which may include naturally occurring substances and reaction/degradation products of fracturing fluid additives. The resulting list of chemicals will be evaluated for existing analytical methods. Where analytical methods exist, detailed information will be compiled on detection limits, interferences, accuracy, and precision. In other instances, standardized analytical methods may not be readily available for use on the types of samples generated by hydraulic fracturing activities. In these situations, a prioritization strategy informed by risk, case studies, and experimental and modeling investigations will be used to develop analytical methods for high-priority chemicals in relevant environmental matrices (e.g., brines).

The sampling and analytical chemistry requirements depend on the specific goals of the field investigation (e.g., detection, quantification, toxicity, fate and transport). Sample types may include formulations of hydraulic fracturing fluid systems, water samples (e.g., ambient water, flowback, and produced water), drilling fluids, soil, and solid residues. In many cases, samples may reflect the presence of multiple phases (gas-liquid-solid) that impact chemical partitioning in the environment. Table H2 briefly discusses the types of analytical instrumentation that can be applied to samples collected during field investigations (both retrospective and prospective case studies).

TABLE H2. OVERVIEW OF ANALYTICAL INSTRUMENTS THAT CAN BE USED TO IDENTIFY AND QUANTIFY CONSTITUENTS ASSOCIATED WITH HYDRAULIC FRACTURING ACTIVITIES

Type of Analyte	Analytical Instrument(s)	MDL Range*
Volatile organics	GC/MS: gas chromatograph/mass spectrometer GC/MS/MS: gas chromatograph/mass spectrometer/ mass spectrometer	0.25-10 µg/L
Water-soluble organics	LC/MS/MS: liquid chromatograph/mass spectrometer/mass spectrometer	0.01-0.025 µg/L
Unknown organic compounds	LC/TOF: liquid chromatograph/time-of-flight mass spectrometer	5 µg/L
Metals, minerals	ICP: inductively coupled plasma	1-100 µg/L
	GFAA: graphite furnace atomic absorption	0.5-1 µg/L
Transition metals, isotopes	ICP/MS: inductively coupled plasma/mass spectrometer	0.5-10 µg/L
Redox-sensitive metal species, oxyanion speciation, thioarsenic speciation, etc.	LC/ICP/MS: liquid chromatograph/inductively coupled plasma/mass spectrometer	0.5-10 µg/L
Ions (charged elements or compounds)	IC: ion chromatograph	0.1-1 mg/L

*The minimum detection limit, which depends on the targeted analyte.

POTENTIAL CHALLENGES

The analysis of field samples collected during case studies is not without challenges. Two anticipated challenges are discussed below: matrix interference and the analysis of unknown chemical compounds.

MATRIX INTERFERENCE

The sample matrix can affect the performance of the analytical methods being used to identify and quantify target analytes; typical problems include interference with the detector signal (suppression or amplification) and reactions with the target analyte, which can reduce the apparent concentration or complicate the extraction process. Some potential matrix interferences are listed in Table H3.

TABLE H3. EXAMPLES OF MATRIX INTERFERENCES THAT CAN COMPLICATE ANALYTICAL APPROACHES USED TO CHARACTERIZE SAMPLES ASSOCIATED WITH HYDRAULIC FRACTURING

Type of Matrix Interference	Example Interferences	Potential Impacts on Chemical Analysis
Chemical	• Inorganics: metals, minerals, ions • Organics: coal, shale, hydrocarbons • Dissolved gases: methane, hydrogen sulfide, carbon dioxide • pH • Oxidation potential	• Complexation or co-precipitation with analyte, impacting extraction efficiency, detection, and recovery • Reaction with analyte changing apparent concentration • Impact on pH, oxidation potential, microbial growth • Impact on solubility, microbial growth
Biological	• Bacterial growth	• Biodegradation of organic compounds, which can change redox potential, or convert electron acceptors (iron, sulfur, nitrogen, metalloids)
Physical	• Pressure and temperature • Dissolved and suspended solids • Geologic matrix	• Changes in chemical equilibria, solubility, and microbial growth • Release of dissolved minerals, sequestration of constituents, and mobilization of minerals, metals

Some gases and organic compounds can partition out of the aqueous phase into a non-aqueous phase (already present or newly formed), depending on their chemical and physical properties. With the numbers and complex nature of additives used in hydraulic fracturing fluids, the chemical composition of each phase depends on partitioning relationships and may depend on the overall composition of the mixture. The unknown partitioning of chemicals to different phases makes it difficult to accurately determine the quantities of target analytes. In order to address this issue, EPA has asked for chemical and physical properties of hydraulic fracturing fluid additives in the request for information sent to the nine hydraulic fracturing service providers.

ANALYSIS OF UNKNOWN CHEMICAL COMPOUNDS

Once injected, hydraulic fracturing fluid additives may maintain their chemical structure, partially or completely decompose, or participate in reactions with the surrounding strata, fluids, gases, or microbes. These reactions may result in the presence of degradates, metabolites, or other transformation products, which may be more or less toxic than the parent compound and consequently increase or decrease the risks associated with hydraulic fracturing formulations. The identification and quantification of these products may be difficult, and can be highly resource intensive and time-consuming. Therefore, the purpose of each chemical analysis will be clearly articulated to ensure that the analyses are planned and performed in a cost-effective manner.

DATA ANALYSIS

The data collected by EPA during retrospective case studies will be used to determine the source and extent of reported drinking water contamination. In these cases, EPA will use different methods to investigate the sources of contamination and the extent to which the contamination has occurred. One important method to determine the source and migration pathways of natural gas is isotopic fingerprinting, which compares both the chemical composition and the isotopic compositions of natural gas. Although natural gas is composed primarily of methane, it can also include ethane, propane, butane, and pentane, depending on how it is formed. Table H4 illustrates different types of gas, the constituents, and the formation process of the natural gas.

TABLE H4. TYPES OF NATURAL GASES, CONSTITUENTS, AND PROCESS OF FORMATION

Type of Natural Gas	Constituents	Process of Formation
Thermogenic gas	Methane, ethane, propane, butane, and pentane	Geologic formation of fossil fuel
Biogenic gas	Methane and ethane	Methane-producing microorganisms chemically break down organic material

Thermogenic light hydrocarbons detected in soil gas typically have a well-defined composition indicative of reservoir composition. Above natural gas reservoirs, methane dominates the light hydrocarbon fraction; above petroleum reservoirs, significant concentrations of ethane, propane, and butane are found (Jones et al., 2000). Also, ethane, propane, and butane are not produced by biological processes in near-surface sediments; only methane and ethylene are products of biodegradation. Thus, elevated levels of methane, ethane, propane, and butane in soil gas indicate thermogenic origin and could serve as tracers for natural gas migration from a reservoir.

The isotopic signature of methane can also be used to delineate the source of natural gas migration in retrospective case studies because it varies with the formation process. Isotopic fingerprinting uses two parameters—δ^{13}C and δD—to identify thermogenic and biogenic methane. These two parameters are equal to the ratio of the isotopes ^{13}C/^{12}C and D/H, respectively. Baldassare and Laughrey (1997), Schoell (1980 and 1983), Kaplan et al. (1997), Rowe and Muehlenbachs (1999), and others have summarized values of δ^{13}C and δD for methane, and their data show that it is often possible to distinguish methane formed from biogenic and thermogenic processes by plotting δ^{13}C versus δD. Thus, the isotopic signature of methane recovered from retrospective case study sites can be compared to the isotopic signature of potential sources of methane near the contaminated site. Isotopic fingerprinting of methane, therefore, could be particularly useful for determining if the methane is of thermogenic origin and in situations where multiple methane sources are present.

In prospective case studies, EPA will use the data collected from field samples to (1) provide a comprehensive picture of drinking water resources during all stages in the hydraulic fracturing water lifecycle and (2) inform hydraulic fracturing models, which may then be used to predict impacts of hydraulic fracturing on drinking water resources.

EVALUATION OF POTENTIAL INDICATORS OF CONTAMINATION

Natural gas is not the only potential chemical indicator for gas migration due to hydraulic fracturing activities: Hydrogen sulfide, hydrogen, and helium may also be used as potential tracers. Hydrogen sulfide is produced during the anaerobic decomposition of organic matter by sulfur bacteria, and can be found in varying amounts in sulfur deposits, volcanic gases, sulfur springs, and unrefined natural gas and petroleum, making it a potential indicator of natural gas migration. Hydrogen gas (H_2) and helium (He) are widely recognized as good fault and fracture indicators because they are chemically inert, physically stable, and highly insoluble in water (Klusman, 1993; Ciotoli et al., 1999 and 2004). For example, H_2 and

He have been observed in soil gas at values up to 430 and 50 parts per million by volume (ppmv) respectively over the San Andreas Fault in California (Jones and Pirkle, 1981), and Wakita et al. (1978) has observed He at a maximum concentration of 350 ppmv along a nitrogen vent in Japan. The presence of He in soil gas is often independent of the oil and gas deposits. However, since He is more soluble in oil than water, it is frequently found at elevated concentrations in soil gas above natural gas and petroleum reservoirs and hence may serve as a natural tracer for gas migration.

EPA will use the data collected from field samples to identify and evaluate other potential indicators of hydraulic fracturing fluid migration into drinking water supplies. For example, flowback and produced water have higher ionic strengths (due to large concentrations of potassium and chloride) than surface waters and shallow ground water and may also have different isotopic compositions of strontium and radium. Although potassium and chloride are often used as indicators of flowback or produced water, they are not considered definitive. However, if the isotopic composition of the flowback or produced water differs significantly from those of nearby drinking water resources, then isotopic ratios could be sensitive indicators of contamination. Recent research by Peterman et al. (2010) lends support for incorporating such analyses into this study. Additionally, DOE NETL is working to determine if stable isotopes can be used to identify Marcellus flowback and produced water when commingled with surface waters or shallow ground water. EPA also plans to use this technique to evaluate contamination scenarios in the retrospective case studies and will coordinate with DOE on this aspect of the research.

References

Baldassare, F. J., & Laughrey, C. D. (1997). Identifying the sources of stray methane by using geochemical and isotopic fingerprinting. *Environmental Geosciences*, *4*, 85-94.

Ciotoli, G., Etiope, G., Guerra, M., & Lombardi, S. (1999). The detection of concealed faults in the Ofanto basin using the correlation between soil-gas fracture surveys. *Tectonophysics*, *299*, 321-332.

Ciotoli, G., Lombardi, S., Morandi, S., & Zarlenga, F. (2004). A multidisciplinary statistical approach to study the relationships between helium leakage and neotectonic activity in a gas province: The Vasto basin, Abruzzo-Molise (central Italy). *The American Association of Petroleum Geologists Bulletin*, *88*, 355-372.

Jones, V. T., & Pirkle, R. J. (1981, March 29-April 3). *Helium and hydrogen soil gas anomalies associated with deep or active faults*. Presented at the American Chemical Society Annual Conference, Atlanta, GA.

Jones, V. T., Matthews, M. D., & Richers, D. M. (2000). Light hydrocarbons for petroleum and gas prospecting. In M. Hale (Ed.), *Handbook of Exploration Geochemistry* (pp. 133-212). Elsevier Science B.V.

Kaplan, I. R., Galperin, Y., Lu, S., & Lee, R. (1997). Forensic environmental geochemistry—Differential of fuel-types, their sources, and release time. *Organic Geochemistry*, *27*, 289-317.

Klusman, R. W. (1993). *Soil gas and related methods for natural resource exploration*. New York, NY: John Wiley & Sons.

Peterman, Z. E., Thamke, J., & Futa, K. (2010, May 14). *Strontium isotope detection of brine contamination of surface water and groundwater in the Williston Basin, northeastern Montana.* Presented at the GeoCanada Annual Conference, Calgary, Alberta, Canada.

Rowe, D., & Muehlenbachs, K. (1999). Isotopic fingerprinting of shallow gases in the western Canadian sedimentary basin—Tools for remediation of leaking heavy oil wells. *Organic Geochemistry, 30,* 861-871.

Schoell, M. (1980). The hydrogen and carbon isotopic composition of methane from natural gases of various origin. *Geochimica et Cosmochimica Acta, 44,* 649-661.

Schoell, M. (1983). Genetic characteristics of natural gases. *American Association of Petroleum Geologists Bulletin, 67,* 2225-2238.

Wakita, H., Fujii, N., Matsuo, S., Notsu, K., Nagao, K., & Takaoka, N. (1978, April 28). Helium spots: Caused by diapiric magma from the upper mantle. *Science, 200*(4340), 430-432.

GLOSSARY

Abandoned well: A well that is no longer in use, whether dry, inoperable, or no longer productive.[1]

ACToR: EPA's online warehouse of all publicly available chemical toxicity data, which can be used to find all publicly available data about potential chemical risks to human health and the environment. ACToR aggregates data from over 500 public sources on over 500,000 environmental chemicals searchable by chemical name, other identifiers, and chemical structure.[15]

Aerobic: Life or processes that require, or are not destroyed by, the presence of oxygen.[2]

Anaerobic: A life or process that occurs in, or is not destroyed by, the absence of oxygen.[2]

Analyte: A substance or chemical constituent being analyzed.[3]

Aquiclude: An impermeable body of rock that may absorb water slowly, but does not transmit it.[4]

Aquifer: An underground geological formation, or group of formations, containing water. A source of ground water for wells and springs.[2]

Aquitard: A geological formation that may contain ground water but is not capable of transmitting significant quantities of it under normal hydraulic gradients.[2]

Assay: A test for a specific chemical, microbe, or effect.[2]

Biocide: Any substance the kills or retards the growth of microorganisms.[5]

Biodegradation: The chemical breakdown of materials under natural conditions.[2]

Casing: Pipe cemented in the well to seal off formation fluids and to keep the hole from caving in.[1]

Coalbed: A geological layer or stratum of coal parallel to the rock stratification.

DSSTox: A public forum for publishing downloadable, structure-searchable, standardized chemical structure files associated with toxicity data.[2]

ExpoCastDB: A database that consolidates observational human exposure data and links with toxicity data, environmental fate data, and chemical manufacture information.[13]

HERO: Database that includes more than 300,000 scientific articles from the peer-reviewed literature used by EPA to develop its Integrated Science Assessments (ISA) that feed into the NAAQS review. It also includes references and data from the Integrated Risk Information System (IRIS), a database that supports critical agency policymaking for chemical regulation. Risk assessments characterize the nature and magnitude of health risks to humans and the ecosystem from pollutants and chemicals in the environment.[14]

HPVIS: Database that provides access to health and environmental effects information obtained through the High Production Volume (HPV) Challenge.

IRIS: A human health assessment program that evaluates risk information on effects that may result from exposure to environmental contaminants. [2]

Flowback water: After the hydraulic fracturing procedure is completed and pressure is released, the direction of fluid flow reverses, and water and excess proppant flow up through the wellbore to the surface. The water that returns to the surface is commonly referred to as "flowback."[6]

Fluid leakoff: The process by which injected fracturing fluid migrates from the created fractures to other areas within the hydrocarbon-containing formation.

Formation: A geological formation is a body of earth material with distinctive and characteristic properties and a degree of homogeneity in its physical properties.[2]

Ground water: The supply of fresh water found beneath the Earth's surface, usually in aquifers, which supply wells and springs. It provides a major source of drinking water.[2]

Horizontal drilling: Drilling a portion of a well horizontally to expose more of the formation surface area to the wellbore.[1]

Hydraulic fracturing: The process of using high pressure to pump fluid, often carrying proppants into subsurface rock formations in order to improve flow into a wellbore.[1]

Hydraulic fracturing water lifecycle: The lifecycle of water in the hydraulic fracturing process, encompassing the acquisition of water, chemical mixing of the fracturing fluid, injection of the fluid into the formation, the production and management of flowback and produced water, and the ultimate treatment and disposal of hydraulic fracturing wastewaters.

Impoundment: A body of water or sludge confined by a dam, dike, floodgate, or other barrier.[2]

Mechanical integrity: An injection well has mechanical integrity if: (1) there is no significant leak in the casing, tubing, or packer (internal mechanical integrity) and (2) there is no significant fluid movement into an underground source of drinking water through vertical channels adjacent to the injection wellbore (external mechanical integrity).[7]

Natural gas or **gas:** A naturally occurring mixture of hydrocarbon and non-hydrocarbon gases in porous formations beneath the Earth's surface, often in association with petroleum. The principal constituent is methane.[1]

Naturally occurring radioactive materials: All radioactive elements found in the environment, including long-lived radioactive elements such as uranium, thorium, and potassium and any of their decay products, such as radium and radon.

Play: A set of oil or gas accumulations sharing similar geologic and geographic properties, such as source rock, hydrocarbon type, and migration pathways.[1]

Produced water: After the drilling and fracturing of the well are completed, water is produced along with the natural gas. Some of this water is returned fracturing fluid and some is natural formation water. These produced waters move back through the wellhead with the gas.[8]

Proppant/propping agent: A granular substance (sand grains, aluminum pellets, or other material) that is carried in suspension by the fracturing fluid and that serves to keep the cracks open when fracturing fluid is withdrawn after a fracture treatment.[9]

Prospective case study: Sites where hydraulic fracturing will occur after the research is initiated. These case studies allow sampling and characterization of the site prior to, and after, water extraction, drilling, hydraulic fracturing fluid injection, flowback, and gas production. The data collected during prospective case studies will allow EPA to evaluate changes in water quality over time and to assess the fate and transport of chemical contaminants.

Public water system: A system for providing the public with water for human consumption (through pipes or other constructed conveyances) that has at least 15 service connections or regularly serves at least 25 individuals.[10]

Redox (reduction-oxidation) reaction: A chemical reaction involving transfer or electrons from one element to another.[3]

Residential well: A pumping well that serves one home or is maintained by a private owner.[5]

Retrospective case study: A study of sites that have had active hydraulic fracturing practices, with a focus on sites with reported instances of drinking water resource contamination or other impacts in areas where hydraulic fracturing has already occurred. These studies will use existing data and possibly field sampling, modeling, and/or parallel laboratory investigations to determine whether reported impacts are due to hydraulic fracturing activities.

Shale: A fine-grained sedimentary rock composed mostly of consolidated clay or mud. Shale is the most frequently occurring sedimentary rock.[9]

Source water: Operators may withdraw water from surface or ground water sources themselves or may purchase it from suppliers.[6]

Subsurface: Earth material (as rock) near but not exposed at the surface of the ground.[11]

Surface water: All water naturally open to the atmosphere (rivers, lakes, reservoirs, ponds, streams, impoundments, seas, estuaries, etc.).[2]

Tight sands: A geological formation consisting of a matrix of typically impermeable, non-porous tight sands.

Toe: The far end of the section that is horizontally drilled.[12]

Total dissolved solids (TDS): All material that passes the standard glass river filter; also called total filterable residue. Term is used to reflect salinity.[2]

ToxCastDB: A database that links biological, metabolic, and cellular pathway data to gene and in vitro assay data for the chemicals screened in the ToxCast HTS assays. Also included in ToxCastDB are human disease and species homology information, which correlate with ToxCast assays that affect specific genetic loci. This information is designed to make it possible to infer the types of human disease associated with exposure to these chemicals.[16]

ToxRefDB: A database that collects *in vivo* animal studies on chemical exposures.[17]

Turbidity: A cloudy condition in water due to suspended silt or organic matter.[2]

Underground injection well (UIC): A steel- and concrete-encased shaft into which hazardous waste is deposited by force and under pressure.[2]

Underground source of drinking water (USDW): An aquifers currently being used as a source of drinking water or capable of supplying a public water system. USDWs have a TDS content of 10,000 milligrams per liter or less, and are not "exempted aquifers."[2]

Vadose zone: The zone between land surface and the water table within which the moisture content is less than saturation (except in the capillary fringe) and pressure is less than atmospheric. Soil pore space also typically contains air or other gases. The capillary fringe is included in the vadose zone.[2]

Water table: The level of ground water.[2]

References

1. Oil and Gas Mineral Services. (2010). *Oil and gas terminology*. Retrieved January 20, 2011, from http://www.mineralweb.com/library/oil-and-gas-terms.
2. US Environmental Protection Agency. (2006). *Terms of environment: Glossary, abbreviations and acronyms*. Retrieved January 20, 2011, from http://www.epa.gov/OCEPAterms/aterms.html.
3. Harris, D. C. (2003). *Quantitative chemical analysis*. Sixth edition. New York, NY: W. H. Freeman and Company.
4. Geology Dictionary. (2006). *Aquiclude*. Retrieved January 30, 2011, from http://www.alcwin.org/Dictionary_Of_Geology_Description-136-A.htm.
5. Webster's New World College Dictionary. (1999). Fourth edition. Cleveland, OH: Macmillan USA.
6. New York State Department of Environmental Conservation. (2011, September). *Supplemental generic environmental impact statement on the oil, gas and solution mining regulatory program (revised draft). Well permit issuance for horizontal drilling and high-volume hydraulic fracturing to develop the Marcellus Shale and other low-permeability gas reservoirs*. Albany, NY: New York State Department of Environmental Conservation, Division of Mineral Resources, Bureau of Oil & Gas Regulation. Retrieved January 20, 2011, from ftp://ftp.dec.state.ny.us/dmn/download/OGdSGEISFull.pdf.

7. U. S. Environmental Protection Agency. (2010). *Glossary of underground injection control terms.* Retrieved January 19, 2011, from http://www.epa.gov/r5water/uic/glossary.htm#ltds.

8. Ground Water Protection Council & ALL Consulting. (2009, April). *Modern shale gas development in the US: A primer.* Prepared for the US Department of Energy, Office of Fossil Energy and National Energy Technology Laboratory. Retrieved January 20, 2011, from http://www.netl.doe.gov/technologies/oil-gas/publications/EPreports/Shale_Gas_Primer_2009.pdf.

9. US Department of the Interior. *Bureau of Ocean Energy Management, Regulation and Enforcement: Offshore minerals management glossary.* Retrieved January 20, 2011, from http://www.mms.gov/glossary/d.htm.

10. U. S. Environmental Protection Agency. (2010.) *Definition of a public water system.* Retrieved January 30, 2011, from http://water.epa.gov/infrastructure/drinkingwater/pws/pwsdef2.cfm.

11. Merriam-Webster's Dictionary. (2011). *Subsurface.* Retrieved January 20, 2011, from http://www.merriam-webster.com/dictionary/subsurface.

12. Society of Petroleum Engineers. (2011). SPE E&P Glossary. Retrieved September 14, 2011, from http://www.spe.org/glossary/wiki/doku.php/welcome#terms_of_use.

13. U.S. Environmental Protection Agency. (2011, September 21). Expocast. Retrieved October 5, 2011, from http://www.epa.gov/ncct/expocast/.

14. U.S. Environmental Protection Agency. (2011, October 31). The HERO Database. Retrieved October 31, 2011, from http://hero.epa.gov/.

15. Judson, R., Richard, A., Dix, D., Houck, K., Elloumi, F., Martin, M., Cathey, T., Transue, T.R., Spencer, R., Wolf, M. (2008) ACTOR - Aggregated Computational Toxicology Resource. Toxicology and Applied Pharmacology, 233: 7-13.

16. Martin, M.T., Judson, R.S., Reif, D.M., Kavlock, R.J., Dix, D.J. (2009) Profiling Chemicals Based on Chronic Toxicity Results from the U.S. EPA ToxRef Database. Environmental Health Perspectives, 117(3):392-9.

17. U.S. Environmental Protection Agency. (2011, October 31). The HERO Database. Retrieved October 31, 2011, from http://actor.epa.gov/actor/faces/ToxCastDB/Home.jsp.